Designing with SOLIDWORKS 2024

Michael J. Rider

SDC
PUBLICATIONS

SDC Publications
P.O. Box 1334
Mission, KS 66222
913-262-2664
www.SDCpublications.com

The Y14 ASME Engineering Drawing and Related Documentation Publications utilized in this text are as follows: ASME Y14.1 1995, ASME Y14.2M-1992 (R1998), ASME Y14.3M-1994 (R1999), ASME Y14.41-2003, ASME Y14.5-1982, ASME Y14.5-1999, and ASME B4.2. Note: By permission of The American Society of Mechanical Engineers, Codes and Standards, New York, NY, USA. All rights reserved.

Download all needed model files from the SDC Publication website (www.SDCpublications.com/downloads/978-1-63057-651-6).

ISBN-13: 978-1-63057-651-6

ISBN-10: 1-63057-651-4

Printed and bound in the United States of America.

Preface

Graphic languages have been around since the dawn of man and, in general, are widely understood by all. A drawing is a graphical representation of a real object, idea, or proposed design for construction at a later time. Engineering drawings or technical drawings are one such form of graphical communication.

This textbook is intended as a classroom text or a self-paced tutorial for the person who wants to learn SOLIDWORKS 2024. Chapter 1 is an introduction to SOLIDWORKS and may be skipped by the reader if they are familiar with SOLIDWORKS. Chapter 2 introduces the reader to the Sketch feature of SOLIDWORKS since it is used to create 3D parts. The meat of this text, learning the basic SOLIDWORKS software, is found in Chapters 3 through 5 and Chapter 7. Chapters 6 and 10 deal with dimensioning and tolerancing an engineering part. Chapters 8 and 9 deal with assemblies and assembly drawings. Chapter 11 is an introduction to SOLIDWORKS SimulationXpress and finite element analysis and may be skipped without loss of course content. A series of appendices are provided to save time for the reader when he/she needs some basic technical information.

Since we are using SOLIDWORKS as a design tool, there are many ways to accomplish the same task. If you are new to SOLIDWORKS, follow the tutorials closely. If you are an experienced SOLIDWORKS user, feel free to deviate from the tutorials occasionally. However, if there are dimensions on a sketch or a 3D part in this textbook, these dimensions are the engineer's design intent. They are important and should be used when creating the part.

All assignments are meant to be printed out on 8.5" x 11" paper. Because it is easier to learn new information if you have a reason for learning it, this textbook discusses design intent while you are learning SOLIDWORKS. The end-of-chapter problems also reference design intent, so looking at the part figure and creating it without reading the design intent mentioned in the description will result in an incorrectly sized part. At the same time, it shows how knowledge covered in basic engineering courses such as statics, dynamics, strength of materials, and design of mechanical components can be applied to design. You do not need an engineering degree or be working toward a degree in engineering to use this textbook. Although FEA (Finite Element Analysis) is used in this textbook, its theory is not covered here.

The author invites all readers of this text to send their comments or suggestions to the author at: Michael J. Rider, Ohio Northern University, Ada, OH 45810 or m-rider.4@onu.edu.

Table of Contents

Chapter 1.   Getting Started 1

     Introduction 1-1

Chapter 2.   Sketch Commands 2-1

     Introduction 2-1

     Sketch Commands Practice 2-11

     Sketch Problems 2-14

Chapter 3.   Extruded Boss/Base 3-1

     Introduction 3-1

     Extruded Boss/Base Practice 3-9

     Extruded Boss/Base Exercise 3-13

     Extruded Boss/Base Problems 3-19

Chapter 4.   Revolved Boss/Base 4-1

     Introduction 4-1

     Revolved Boss/Bass Practice 4-7

     Revolve Boss/Base Exercise 4-13

     Revolved Boss/Base Problems 4-20

Chapter 5.   Patterned and Mirrored Features 5-1

     Introduction 5-1

     Patterned Features Practice (Linear) 5-7

     Patterned Features Practice (Circular) 5-13

     Patterned Features Exercise (Circular) 5-19

     Patterned Features Problems 5-28

Chapter 6.   Dimensioning 6-1

     Introduction 6-1

     Learn to Dimension Properly 6-2

     Create a Custom Drawing Template 6-28

     Creating a Detailed Engineering Drawing 6-37

     Creating a Sectioned View 6-45

     Creating an Auxiliary View 6-48

     Dimensioning Problems 6-50

Chapter 7. Parametric Modeling using Equations 7-1

 Introduction 7-1

 Design Intent, Using Equations, and Patterns...................................... 7-9

 Parametric Modeling Using Equations Problems 7-20

Chapter 8. Assemblies and Subassemblies... 8-1

 Introduction 8-1

 Assembly Practice... 8-3

 Assembly Exercise... 8-13

 Assembly Problems 8-35

Chapter 9. Assembly Drawing... 9-1

 Introduction 9-1

 Assembly Drawing Practice.. 9-4

 Assembly Drawing Exercise... 9-8

 Assembly Drawing Problems 9-11

Chapter 10. Tolerancing and GD&T 10-1

 Introduction 10-1

 Tolerancing and GD&T Practice 10-13

 Tolerancing and GD&T Exercise 10-21

 Tolerancing and GD&T Problems 10-25

Chapter 11. Introduction to Finite Element Analysis 11-1

 Introduction 11-1

 Finite Element Analysis Practice 11-4

 Finite Element Analysis Exercise 11-9

 Finite Element Analysis Problems.. 11-14

Chapter 12. Appendices.. 12-1

 Appendix A – Drill and Tap Chart 12-2

 Appendix B – Number and Letter Drill Sizes...................................... 12-4

 Appendix C – Surface Roughness Chart 12-5

 Appendix D – Clevis Pin Sizes.. 12-6

 Appendix E – Square and Flat Key Sizes 12-7

 Appendix F – Screw Sizes 12-9

 Appendix G – Nut Sizes 12-11

 Appendix H – Setscrew Sizes 12-12

 Appendix I – Washer Sizes.. 12-13

Appendix J – Retaining Ring Sizes 12-15

Appendix J – Basic Hole Tolerance.. 12-17

Appendix K – Basic Shaft Tolerance... 12-18

Appendix L – Inch Tolerance Zones 12-19

Appendix M – Metric Tolerance Zones.. 12-20

Appendix N – International Tolerance Grades 12-21

Chapter 13.   References... 13-1

Chapter 1.   Getting Started

Introduction

This book will concentrate on the modeling application used to create solid models of parts, engineering design and drawings, and assemblies.

The tutorials in this book assume you are using the default settings of SOLIDWORKS in a university setting.

SOLIDWORKS uses the mouse and its three buttons extensively. Therefore, it is essential to understand the mouse button's basic functions listed below.

LMB—(LEFT MOUSE BUTTON)

Used for most operations such as selecting icons, picking graphic entities, and selecting menu items.

MMB—(MIDDLE MOUSE WHEEL)

Used to maneuver the entities in the window. To **Rotate** hold down the wheel button then move the mouse. To **Pan** hold down the wheel button along with the <Ctrl> key and move the mouse. To **Zoom** hold down the wheel button along with the <Shift> key and move the mouse. Spinning the wheel causes the entities to zoom in or out relative to the position of the mouse cursor.

RMB—(RIGHT MOUSE BUTTON)

Hold down the button to bring up additional available options in a context-sensitive pop-up menu.

The **<Esc>** key is used to cancel a command. Sometimes it may be necessary to press the **<Esc>** key twice to cancel a command.

The **<F>** key resizes the entities in the window to fit in the window. Entities typically go off-screen when they are being resized or moved.

The **<G>** key activates the magnifying glass feature so you can zoom in on an area.

<Z> key and Shift+<Z> key

Pressing the <Z> key zooms out. Holding down the <Shift> key and pressing the <Z> key zooms in.

Using just the Arrow keys

Pressing the arrow key rotates the entities on the screen left, up, down, or right.

Holding down the <Shift> key and pressing the arrow keys rotate the entities on the screen 90 degrees left, up, down, or right.

Holding down the <Alt> key and pressing the left arrow key rotates the entities counterclockwise. <Alt> key and right arrow key rotate the entities clockwise and normal to the screen.

Below is a list of other keystrokes used in SOLIDWORKS and their meaning.	
<Ctrl> + <1>	makes the Front View normal to the screen
<Ctrl> + <2>	makes the Back View normal to the screen
<Ctrl> + <3>	makes the Left View normal to the screen
<Ctrl> + <4>	makes the Right View normal to the screen
<Ctrl> + <5>	makes the Top View normal to the screen
<Ctrl> + <6>	makes the Bottom View normal to the screen
<Ctrl> + <7>	makes the Isometric View normal to the screen
<Ctrl> + <8>	makes the current selected surface normal to the screen

Note: It is good practice to create a **SOLIDWORKS Template** folder to store your templates in and a **SOLIDWORKS Parts** folder to store your data files in before starting SOLIDWORKS. When saving files in SOLIDWORKS, you should always verify that you are saving them in the proper folder before picking the Save button.

When working with SOLIDWORKS, it is strongly advised to **_save your work often_** in case of a power failure or an unexpected program crash.

> **Note that SOLIDWORKS 2024 is now backward compatible with SOLIDWORKS 2023 and 2022. SOLIDWORKS 2023 and 2022 were not backward compatible with older versions.**

1. Before starting SOLIDWORKS, create a **SOLIDWORKS Part** folder and a **SOLIDWORKS Template** folder on the desktop.

2. 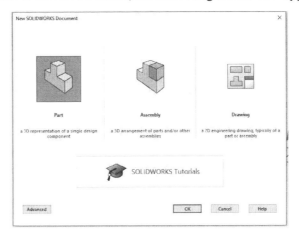 Double-click the left mouse button (LMB) on the SOLIDWORKS 2024 icon on the desktop to start SOLIDWORKS.

3. If SOLIDWORKS is in Novice mode, the following screen will appear.

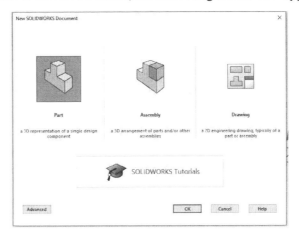

4. Pick the Part icon, then pick the OK button. The following screen will appear.

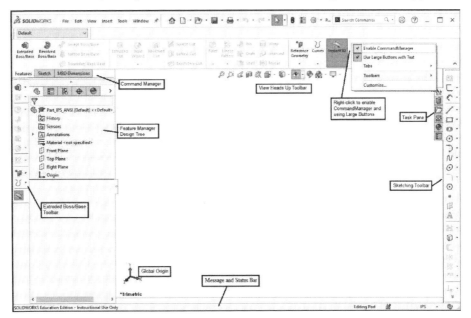

5. ⚙ Pick the Options icon at the top of the window. When the System Options window appears, pick Color. Then select Viewport Background in the Color Scheme settings area. Pick the Edit button, then select White from the color menu. Change the Top Gradient Color and the Bottom Gradient Color to white as well.

6. Select the **Plain (Viewport Background color above)** option. Pick the OK button. This will produce a white background while you are in SOLIDWORKS, which is my preferred background color.

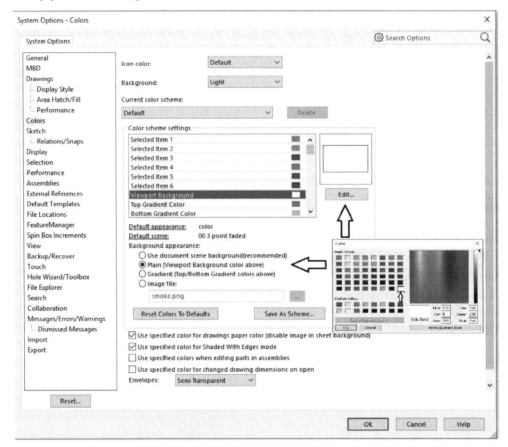

7. In the lower right corner of the screen, pick the unit's area (either IPS or MMGS), then pick Edit Document Units… Most of the exercises in this textbook use the IPS system so make sure IPS is selected.

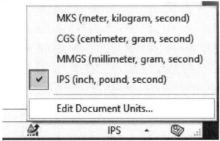

8. Pick the Document Properties tab. Set Drafting Standards to ANSI and Length Units to 3-decimal places for inches, then pick **OK**.

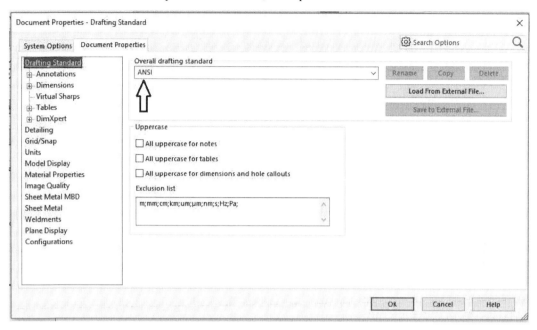

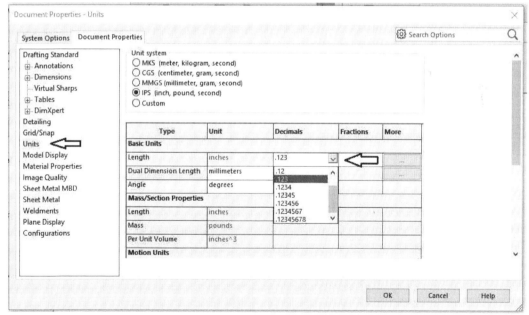

9.   Under File at the top of the screen, pick **Save As…**

10. For the Save as type: select "**Part Template (*.prtdot)**" from the pulldown menu. Enter "**Part_IPS_ANSI**" as the filename. At the top of the window navigate to the **SOLIDWORKS Template** folder you created previously, and pick the **Save** button. If a file called "**Part.PRTDOT**" is in the same directory, you can delete it.

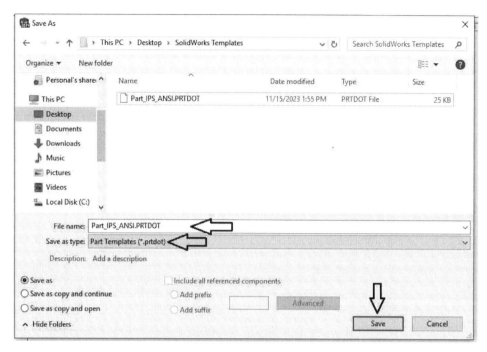

11. Under File, pick **Close**.

Now we need to add this folder to file locations that SOLIDWORKS can look into.

12. Under the Tools menu, pick **Options**. Select **File Locations** under System Options. Make sure **Document Templates** is selected as the **Show folders for:** option. Pick the **Add...** button to add the user-define template to the list of folders containing document templates. Locate the SOLIDWORKS Template folder on the Desktop, then pick the **Select Folder** button.

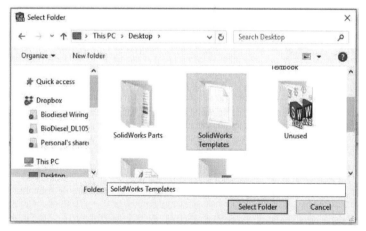

13. Pick the **OK** button. If a pop-up window appears with the question, "Would you like to make the following changes to your search paths?" Pick **Yes**.

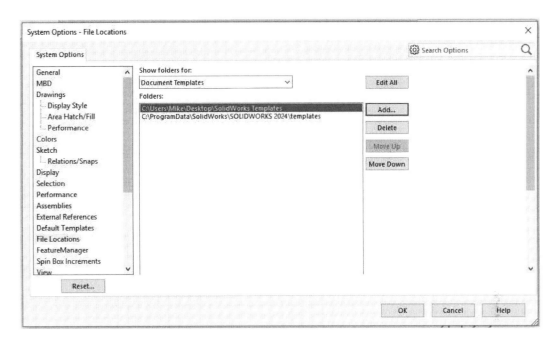

We will use this template for each new part we create so that we don't have to set these parameters each time we create a new part.

14. Under File, pick **New…** Pick the **Advanced** button in the lower-left corner of the window if SOLIDWORKS comes up in Novice mode.

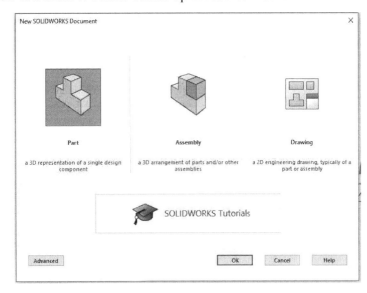

15. Pick the SOLIDWORKS Templates tab, then pick the **Part_IPS_ANSI** icon. Pick OK.

We are back in SOLIDWORKS with all desired parameters already set. If the left
Extruded Boss/Base Toolbar and the right Sketching Toolbar are not present, don't
worry; they will be added shortly.

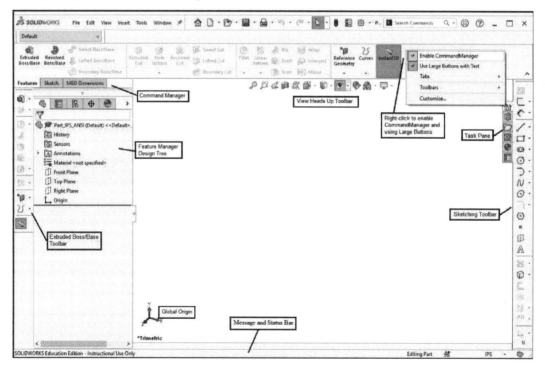

Note that I only have 3 tabs (Features, Sketch, and MBD Dimensions) listed in my working
window because I have turned off the other tabs. Right-click at the top and select the **Tab**
option to turn off the tabs you do not want to see or use.

With **Use Large Buttons with Text** selected, the three tabs appear as follows.

The Sketch tab.

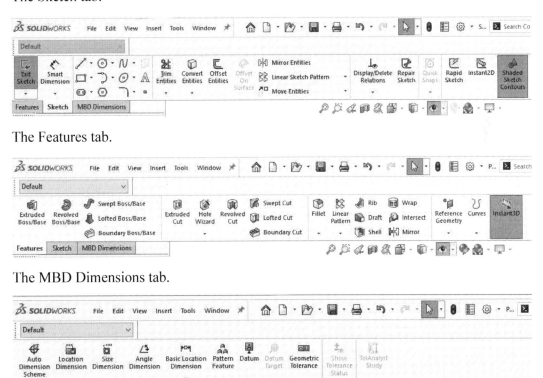

The Features tab.

The MBD Dimensions tab.

You can also add the Sketch and Features tools along the side of the window. The Sketch tools will appear along the right side of the window. The Extruded Boss/Base Toolbar will appear along the left side of the window. These are added by **Toolbars** from the same pop-up menu.

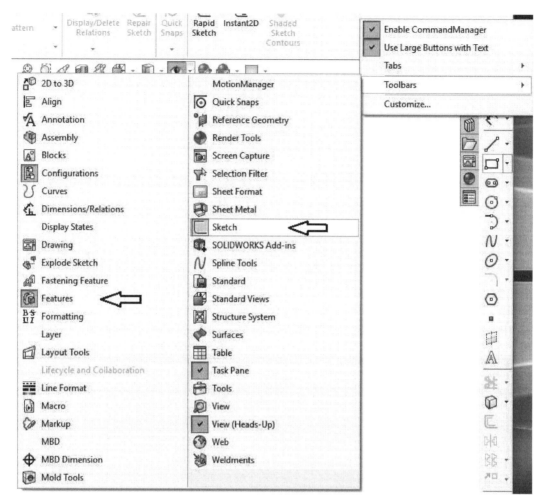

Selecting the **CommandManager** from the pop-up menu disables it. Picking it again enables it. Disabling the **Use Large Buttons with Text** option gives you more room to work so do this now.

16. To start a sketch, first select one of the three principle planes (Front Plane, Top Plane, or Right Plane), then pick the Sketch icon at the top of the window or along the right side of the window using the LMB. Pick the **Front Plane** for this exercise.

If you forget to pick the plane first, the three orthogonal planes will appear on the screen so you can select the desired sketching plane such as the **Front Plane**.

Before sketching it is a good idea to turn on **View Sketch Relations** so you can tell when a line is horizontal, vertical, parallel to another line, perpendicular to another line, tangent to a curve, etc.

17. Pick the **eyeball** icon in the **View Heads Up Toolbar**. Verify that the **View Sketch Relations** icon is highlighted, otherwise select it.

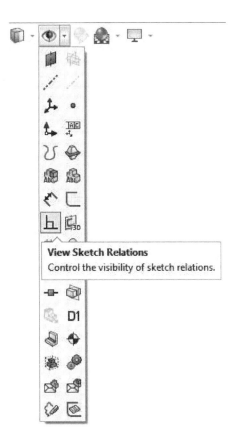

18. 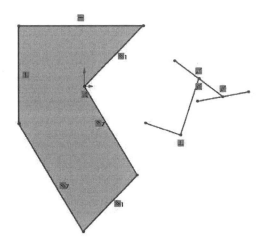 Select the **Line** tool, then draw a few lines similar to what is shown below. The exact locations or lengths of the lines don't matter. Typically, you want to start at the plane's origin. If you create a closed area, the area will be shaded as shown below and the line command will be paused for one mouse click. To stop a chain of connected lines, double-click the left mouse button (LMB). The line command is still active so you can continue to draw additional lines. To stop drawing lines, press the <ESC> key, or Right-click (RMB) and pick **Select** from the pop-up menu.

If the sketch relations do not show up on your sketch, see step 17.

In the figure above, the **Coincident** Sketch Relations icon indicates that the first point is coincident with the Front Plane's origin. There are two sets of parallel lines. There is a vertical and a horizontal line, and two lines that are perpendicular to each other. On the right, one line connects at the midpoint of another line and is coincident with that line. This information can be very helpful when sketching a given shape. A list of Sketch Relations is shown below.

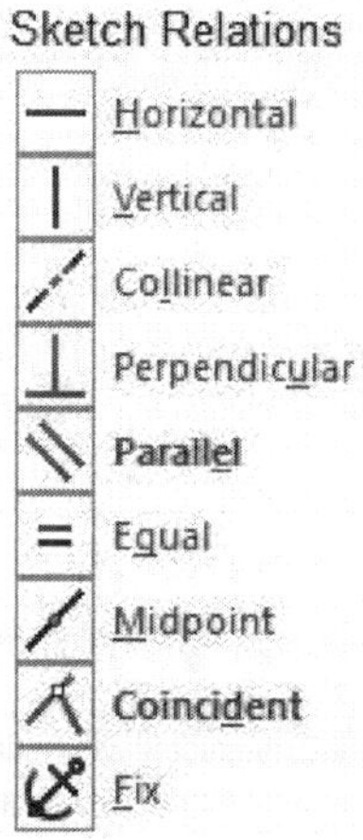

Sketch Relations

—	Horizontal	**Horizontal** forces the selected lines to be horizontal
│	Vertical	**Vertical** forces the selected lines to be vertical
╱	Collinear	**Collinear** for the selected lines to be in line with each other
⊥	Perpendicular	**Perpendicular** forces two lines to be perpendicular
╲	Parallel	**Parallel** forces the selected lines to be perpendicular
=	Equal	**Equal** force the selected objects to be equal in size
╱	Midpoint	**Midpoint** forces the selected object to connect at the midpoint
⅄	Coincident	**Coincident** forces two points or objects to be connected
⊗	Fix	**Fix** forces the selected object to be fixed in the window

If your sketched lines do not reflect the desired Sketch Relations, you can select the appropriate lines by picking the first line, then holding down the <CTRL> key and picking the second line. A list of possible sketch relations will appear on the left side of the window. Pick the appropriate sketch relation from the list.

19. Move to the upper portion of the sketching window and hold the LMB down. Move the mouse to the lower right area of the window, which will highlight all of the lines previously drawn. Release the LMB, then press the <Delete> key to delete all of the selected lines.

20.  ⊙ Select the Circle icon from the sketch toolbar. Pick a point in the window, then move the mouse to a new location. A Circle's center will be located at the first point selected and the size will be dictated by the second mouse pick. Draw several circles similar to what is shown below.

21. Pick the center of the circle at the origin, hold down the <Ctrl> key, and pick the center of the smaller circle below it. In the Add Relations window, pick the Vertical icon. Pick the green checkmark.

22. Pick the center of the left circle, then hold down the <Ctrl> key, and pick the center of the right circle. In the Add Relations window, pick the Horizontal icon. Pick the green checkmark.

23. Pick the edge of the left circle, then hold down the <Ctrl> key, and pick the edge of the right circle. In the Add Relations window, pick the Equal icon. Pick the green checkmark.

24. Pick the edge of the left circle, then hold down the <Ctrl> key, and pick the edge of the circle centered at the origin. In the Add Relations window, pick the Tangent icon. Pick the green checkmark.

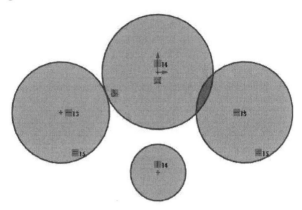

The first circle was drawn at the origin of the sketch window. The smaller circle is directly below the first circle as indicated by the Vertical relation. The left and right circles are equal in size and their centers are along the same horizontal axes.

25. Select the Line tool again, then add four lines that are tangent to the four circles as shown. If any of the lines are not tangent to the given circle, select the circle near the endpoint of the line, then pick Tangent from the Sketch Relations window on the left. If another relation is present other than the coincident relation at the end of a given line, pick the relation with the LMB, then press the <Delete> key to remove it. Each line segment should have a Tangent and Coincident relation attached. See the figure below.

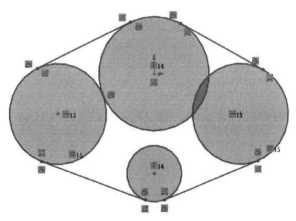

Use the **Trim Entities** tool to remove unwanted portions of your sketch.

26. Select the **Trim Entities** icon. Move the mouse inside the four tangent lines, then press down and hold the LMB, and drag the mouse cursor around without touching any of the four lines or the outside region of the four circles. Your sketch should look similar to the figure below. If you accidentally touch a line or the outer portion of a circle, undo the changes by pressing <Ctrl>+Z and try again.

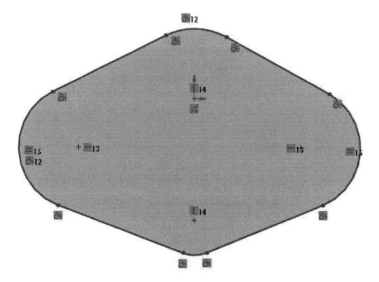

At the bottom of the window, the phrase Under Defined is visible. This is because we haven't sized any of the circles or located the centers of any of the circles. We will do this now. Also, note that the sketch is not currently symmetrical.

27. 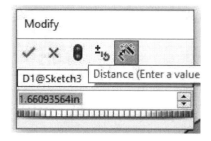 Pick the **Smart Dimension** icon. Pick the left circle, then move away from it and pick the location for the size dimension. Type 1.5 for the value, then press the <Enter> key or pick the green checkmark in the pop-up window.

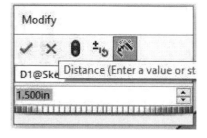

The radius of the left arc will be resized to 1.50 inches. Note that the right arc will also be 1.50 inches since the two arcs were set equal in size previously.

28. Pick the top arc and set its size to 1.75 inches.
29. Pick the bottom arc and set its size to 1.00 inches.
30. Pick the center of the left arc, then the center of the right arc. Move the mouse cursor down and note that a dimension appears between the centers of the two arcs. Set its value to 5.50 inches.
31. Set the distance between the top arc's center and the bottom arc's center equal to 3.00 inches.

At the bottom of the window, the phrase Under Defined is still showing so there is at least one more dimension or sketch relation missing. Let's make the object symmetrical. To do this, we need to add a centerline, and then add a symmetric relation.

32. 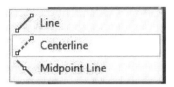 Pick the down arrow to the right of the Line icon to bring up a pop-up menu. Pick Centerline from this menu, then draw a vertical centerline through the origin.

33. Pick the center of the left arc. Hold down the <Ctrl> key and pick the center of the right arc, then pick the centerline. Select **Symmetric** from the Add Relation window. (You can pick these three entities in any order.)

At the bottom of the window, the phrase **Fully Defined** appears, and your sketch should look exactly like the sketch below.

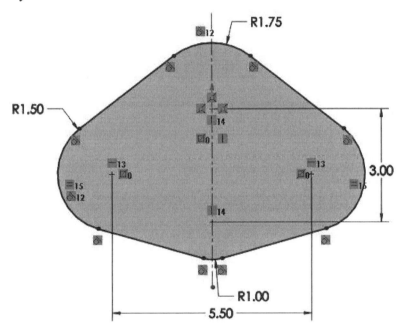

Let's exit from the sketch, and extrude this shape to make our first 3D part with a depth of 2.00 inches.

34. Pick the shaded **Sketch** icon on the right or the **Exit Sketch** icon at the top of the screen if the **Use Large Buttons with Text** feature is enabled.

35. With the sketch highlighted in the **Feature Manager Design Tree**, pick the Extruded Boss/Base icon. Leave the direction as Blind. Set the depth in Direction 1 to 2.00 inches. Pick the green checkmark to complete the extrusion.

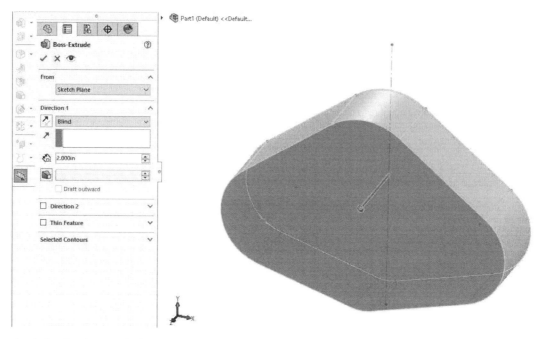

And the final extruded part appears on the screen.

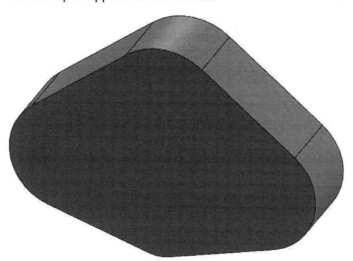

36. If necessary, press the <F> key to fit the part on the screen.
37. Right-click on the **Material** icon in the **FeatureManager Design Tree** and select **Plain Carbon Steel** using the LMB. Notice that the selected material now appears in the **FeatureManager Design Tree**.

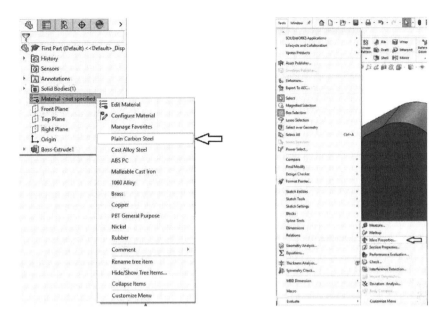

38. Under the File menu, pick **Save**. Name the part as "First Part". Make sure you are saving the part file in the **SOLIDWORKS Parts** folder you created previously. Pick **Save**.

39. Under **Tools**, select **Evaluate**, then select **Mass properties...** The following window appears. The mass should be 18.88 pounds and the volume should be 67.01 cubic inches if you properly dimensioned the part and added the appropriate sketch relations. If not, go back and check your construction procedure.

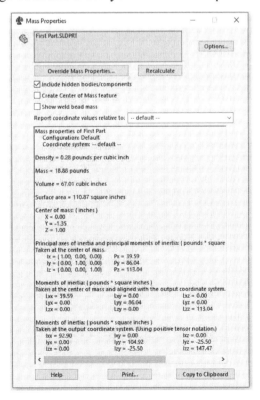

40. Pick the X in the upper right corner of the window to close the Mass Properties window.

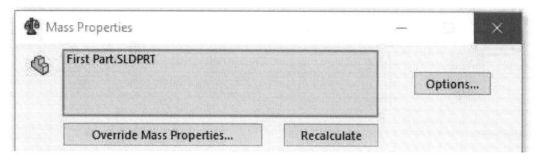

41. If your mass or volume is different, right-click on **Boss-Extrude1** and select **Edit Sketch**, then check over your sketch for this part. Maybe you are missing a sketch relation. After correcting your sketch, exit your sketch, and check **Mass properties...** again.

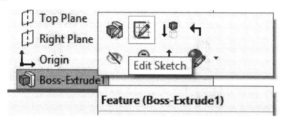

42. 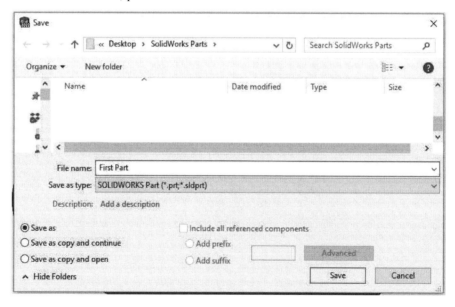 Pick **Save**.
43. Under the File menu again, pick **Close**.
44. Under the File menu, pick **Exit** to exit from SOLIDWORKS.

Chapter 2.   Sketch Commands

Introduction

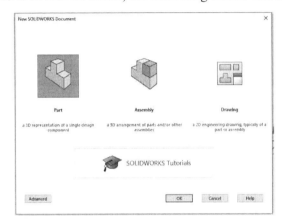

1. 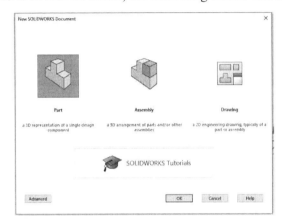 Double-click on the SOLIDWORKS 2024 icon on the desktop to start SOLIDWORKS using the left mouse button (LMB).
2. If SOLIDWORKS is in Novice mode, the following screen will appear.

3. If in Notice mode, pick the Advance button.
4. Under File, pick **New…**
5. Pick the SOLIDWORKS Templates tab, then pick the **Part_IPS_ANSI** icon. Pick OK. The following screen with appear.

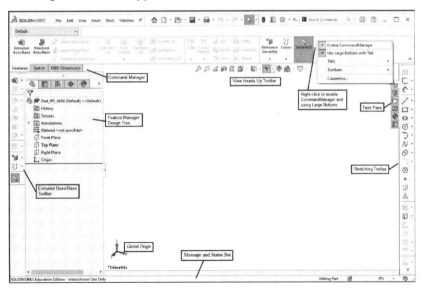

6. If the Sketch and Features tools are not on the sides of the window, you can also add the Sketch and Features tools to the sides of the window. The Sketch tools will appear along the right side of the window. The Features tool will appear along the left side of the window. These are added using the **Toolbars** option from the same pop-up menu.

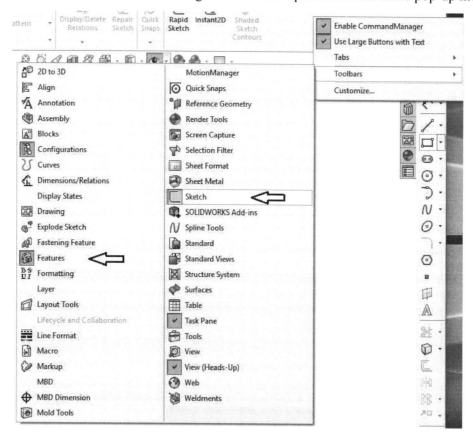

7. Selecting the **CommandManager** from the pop-up menu disables it. Picking it again enables it. Disabling the **Use Large Buttons with Text** option gives you more room to work.

    Let's go on with learning more about sketching features.

8. To start a sketch, first select one of the three principle planes (Front Plane, Top Plane, or Right Plane), then pick the **Sketch** icon at the top of the window or along the right side of the window using the LMB. Pick the **Front Plane** for this exercise.

9. Select the **Line** tool, then draw two parallel lines at some angle similar to what is shown below. The exact locations or lengths of the lines don't matter. Typically, you want to start at the plane's origin so draw one line starting at the origin of the Front plane.

10. ◫ If the **Parallel** icon doesn't show up on the two lines, select one line, hold down the <Ctrl> key, and pick the second line. In the Add Relations window pick the **Parallel** icon.

11. ◫ Select the **Tangent Arc** icon, then select the end of one of the lines, then move away from the line and toward the second line's endpoint. When it connects to the second line, press the LMB.

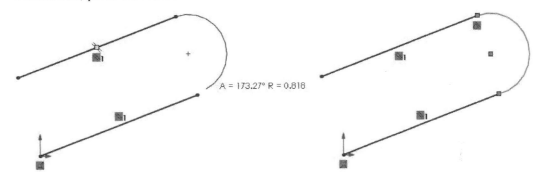

12. Pick the end of one of the lines, then pick a point on the previously drawn arc as the second point. Note that arcs can also end at a random location or another arc, but they must begin at an endpoint.

13. ◫ Select the **Centerpoint Arc** icon, then pick the end of one of the lines as the center point of the arc. Move away from the point and a dashed line and arc will appear. Use the LMB to pick the starting point of the arc. Note that you can only move along the arc's path after this. Move to a new location and press the LMB again to end the arc. Press the <Esc> key or RMB and pick **Select** from the pop-up menu to exit from this command.

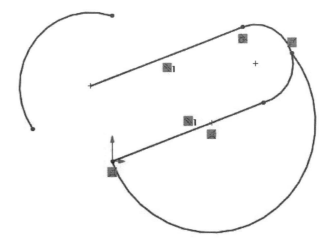

14. Erase the entire sketch by typing <Ctrl>+A followed by <Delete>.

15. ◫ Select the **Corner Rectangle** icon, then draw several rectangles. Note that all of the lines are either horizontal or vertical as indicated by the sketch relation icons.

16. Select the **Center Rectangle** from the pop-up menu for rectangles. Pick a point as the center of the rectangle, then move the mouse cursor away and watch the rectangle grow in both directions. Use the LMB to complete the rectangle.

17. Select the **3-Point Center Rectangle** icon. Pick a point for the center of the rectangle. Move the cursor away at an angle and pick a second point as the direction for the rectangle. Finally, move the cursor perpendicular to the line just drawn and use the LMB to size the rectangle in the second direction.

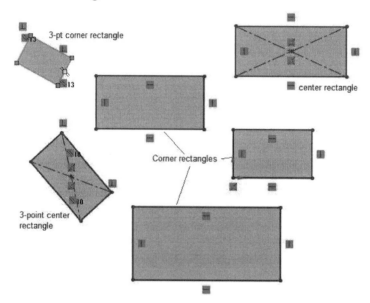

18. Select the **3-Point Corner Rectangle** icon. Pick a point as the first corner of the rectangle. Move away from the first point creating one side of the rectangle. Press the LMB to set the second corner. Now the cursor will only move perpendicular to the dashed line drawn to size the rectangle in the other direction. Press the LMB to size the rectangle based on its three corners.

19. Press <Esc> to exit from the previously activated drawing command. Or Right-click to bring up a pop-up menu, then pick the Select option.

20. Type <Ctrl>+A to select all drawing features, then press the <Delete> key to erase the entire sketch.

Starting over with a blank sketch window, let's draw some slots.

21. Select the **Straight Slot** icon. Pick a point that represents the center of one of the end arcs. Move the cursor to determine the direction of the slot and press the LMB to locate the center of the second arc. Move the cursor perpendicular to this dashed line to size the arcs.

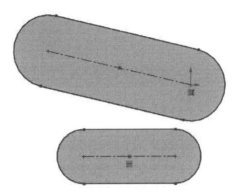

22. 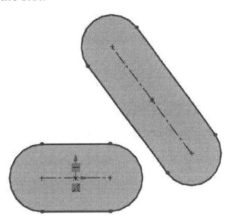 Select the **Centerpoint Straight Slot** icon. Pick a point as the center of the slot. Move away in the direction of the slot, then press the LMB to determine the length of the slot. Finally, move perpendicular to this dashed line and press the LMB to determine the width of the slot.

23. Erase the entire sketch by typing <Ctrl>+A followed by <Delete>.

24. 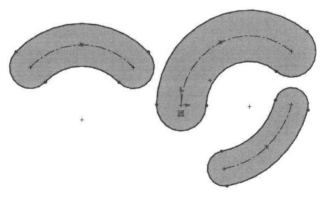 Select the **3-Point Arc Slot** icon. Pick the center point for one of the slot's arcs. Move the cursor and pick the center for the other slot's arc. Move the cursor to size the slot's arc curvature. Pick a fourth point to size the slot's width.

25. Erase the entire sketch by typing <Ctrl>+A followed by <Delete>.

26. Select the **Centerpoint Arc Slot** icon. Select the center point for the arc. Use the LMB to pick one end of the slotted arc. Use the LMB to pick the other end of the slotted arc. Finally use the LMB to set the width of the slotted arc.

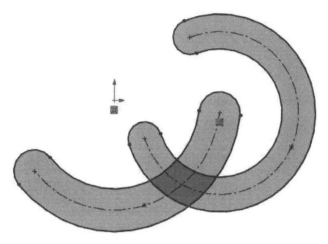

27. Erase the entire sketch by typing <Ctrl>+A followed by <Delete>.

28. Select the **Corner Rectangle**, then draw a rectangle of any given size with one corner at the origin. Press <Esc>, then select the **Sketch Fillet** icon to round a corner of the rectangle. Use the LMB to pick the two edge lines of the rectangle to place the rounded corner at their intersection point. The default size is typically 0.1 inches. In the Fillet Parameters window, change this value to 0.25 inches, then pick the green checkmark. Although you can add fillets to your sketch, it is better to add them to the 3D Part.

29. Use the **Smart Dimension** tool to verify the size of the fillet if it doesn't show up automatically.

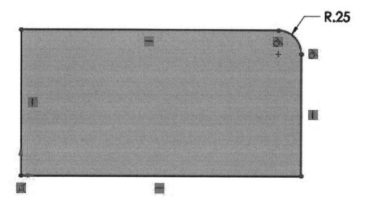

Although you can add chamfers to your sketch, it is better to add them to the 3D Part. The **Sketch Chamfer** tool is shown here but typically is not added to a sketch. To use it, pick the icon, then pick the two edges where the chamfer is to be placed, then enter the size of the chamfer.

Sketch Chamfer places a chamfer of a given leg size and angle at the corner specified, or the size of the two legs of the chamfer at the corner specified. You can typically select Angle-distance or Distance-distance; however, only Distance-distance seems to be working in SOLIDWORKS 2024 SP0.1

30. Select the Polygon icon to draw a polygon consisting of three or more sides. Pick the location for the center of the polygon, then move the cursor away to set the size of the polygon. Note you can also adjust the initial angle for the polygon by the direction you move the cursor. After you press the LMB, the Polygon window appears which allows you to set the number of sides in the polygon along with whether there is an inscribed or circumscribed circle associated with the polygon. Pick the green checkmark to finalize the polygon's initial size and placement.

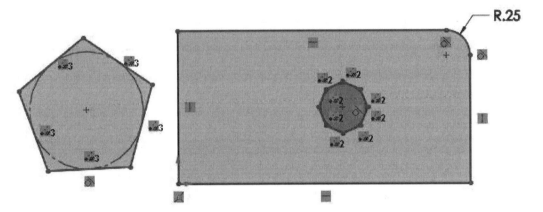

31. Move the cursor to the upper left area of the window, press and hold the LMB, and drag the cursor to the lower right area of the window, which will select all entities in this region. Press the <Delete> key to erase the area selected.

The **Linear Sketch Pattern** command is used to duplicate entities in one-dimensional or two-dimensional directions.

32. Select the **Corner Rectangle** then draw a rectangle at the origin.

33. Select the **Smart Dimension** icon. Set the width to 6 inches and its height to 4 inches.

34. Select the **Circle** icon and draw a circle near the lower left corner of the rectangle. Use the **Smart dimension** tool to locate it 0.75 inches from the left and bottom of the rectangle. Set its diameter to 0.25 inches.

35. Select the **Linear Sketch Pattern** icon. Set the X-axis spacing to 0.5 inches and the number of incidents to 10. Set the number of Y-axis incidents to 6 and the spacing to 0.5 inches. Check the Dimension X spacing box and the Dimension Y spacing box. Uncheck the Display instance count boxes for both directions. Select the **Entities to**

Pattern area, then pick the circle just drawn. Pick the green checkmark to process the command. Imagine the time savings when sketching this part with sixty evenly-spaced holes.

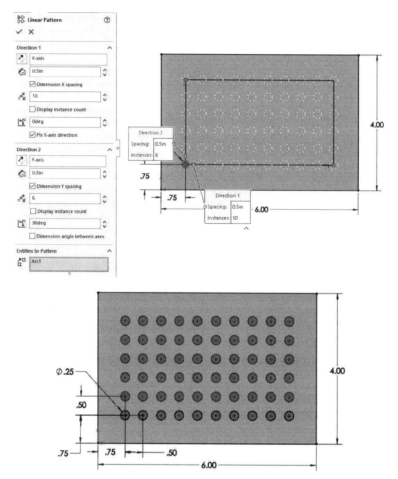

Instead of using the X-axis and Y-axis directions, this tool can duplicate entities along a selected line in either direction.

36. Erase the entire sketch by typing <Ctrl>+A followed by <Delete>.

37. Select the Corner Rectangle icon then draw a rectangle with its lower left corner at the plane's origin. Do not add any dimensions.

> **Note: You can fully define the sketch using the option in steps 38, 39, and 40.**

38. **Since your sketch is not Fully Defined, you can select the** Fully Define **Sketch** option under the **Display/Delete Relations** icon.

39. Make sure **All entities in sketch** is selected, and the **Relations** and **Dimensions** boxes are checked.

40. Pick the **Calculate** button. SOLIDWORKS will add the dimensions and relations to make the sketch fully defined. Pick the green checkmark.

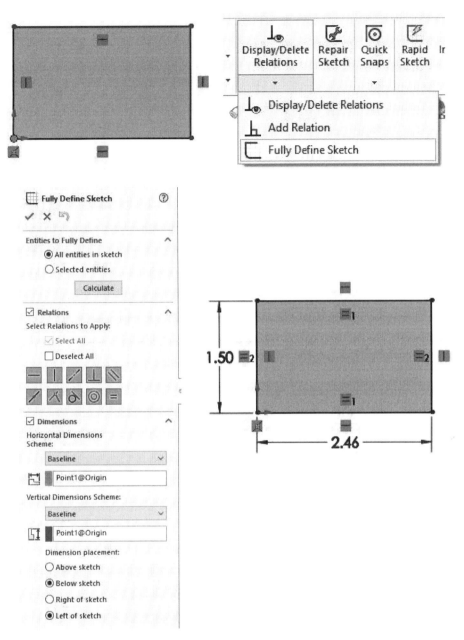

The **Construction Geometry** icon changes a sketch element into a construction entity or a construction entity back into a sketch element.

41. Select the top line of the rectangle with the LMB.

42. ⇄ Select the **Construction Geometry** icon from the pop-up menu.

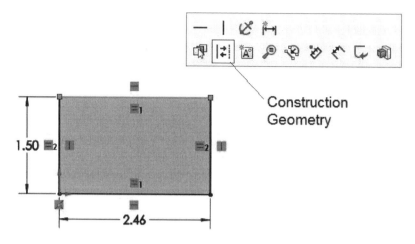

Construction
Geometry

The top line is now a construction line.

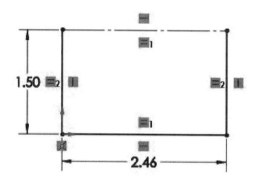

43. Select the top line again, then pick the **Construction Geometry** icon and the line changes back into a sketch element.

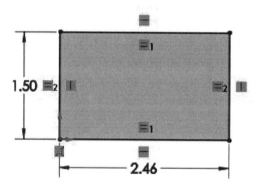

44. Erase the entire sketch by typing <Ctrl>+A followed by <Delete>.

Sketch Commands Practice

Let's practice a few of the sketch tools by creating the sketch shown here.

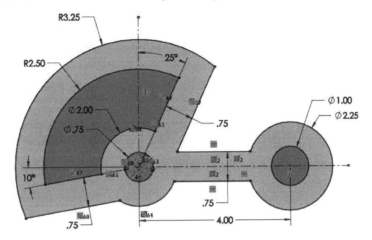

1. 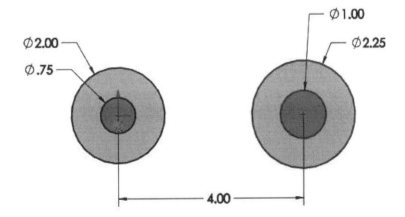 Draw two coaxial circles at the origin of the Front plane. Draw two additional circles to the right of the first two circles.

2. Pick the center of the first set of two circles, hold down the <Ctrl> key, and pick the center of the second set of two circles. Pick Horizontal from the Sketch Relations window.

3. Use the **Smart Dimension** tool to size the four circles and the distance between their centers.

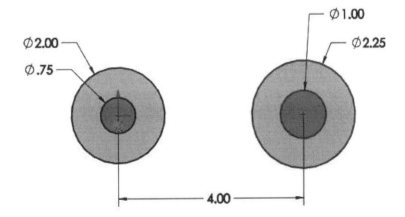

4. Draw a horizontal centerline through the origin and the centers of the two sets of circles.

5. Use the **3-Point Corner Rectangle** tool to draw a rectangle coincident with the outer two circles and approximately symmetrical about the horizontal centerline.

6. Use the **Trim Entities** tool to remove the ends of the rectangle. The entire area should be shaded at this point. If not, there is probably a small gap between features. Make sure the two horizontal lines are coincident with (connected to) the outer arcs and symmetrical about the horizontal centerline.

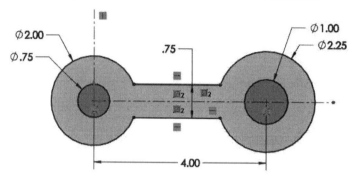

7. Draw a vertical centerline through the origin.

8. Starting at the origin use the **Line** tool to draw a line at a 25-degree angle from the vertical centerline and a second line 10 degrees below the horizontal centerline as shown below.

9. Use the **Smart Dimension** tool to correctly orient these line segments.

10. Use the **Centerpoint Arc** tool to draw a 2.50-inch arc between the two angled lines. Trim off any excess length on either line.

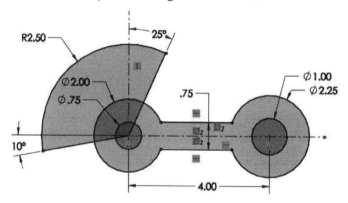

11. Draw a line parallel to the 25-degree line approximately ¾ inch away starting at the horizontal centerline. Draw a line parallel to the 10-degree line approximately ¾ inch below starting at the vertical centerline. Afterward, pick the two almost parallel lines and make them parallel.

12. Use the **Centerpoint Arc** tool to draw a second arc between the two newly drawn lines.

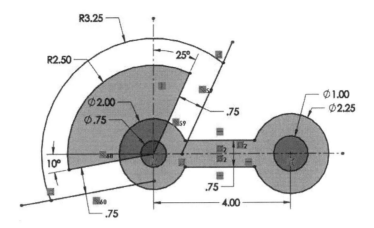

13. Use the Trim Entities tool to remove the line segments and arc segments necessary to create a closed region. See the figure below.

14. Use the Smart Dimension tool to size the thickness of the angled section at 0.75 inches

Your sketch should appear as shown below. All the lines should be black and the phrase "Fully Defined" should appear at the bottom of the sketch window. If any line is blue, then you are missing a dimension or a sketch relation. If the enclosed area is not shaded, then you may have overlapping entities or a gap between entities.

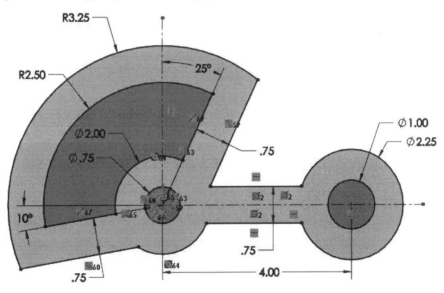

15. File> Save. Enter the name **"Adjustable Sector"**. Make sure you are saving the part file in the **SOLIDWORKS Parts** folder. Pick the **Save** button.

16. **File> Print…** (if asked to)

17. **File> Close**

Sketch Problems

2-1   Draw the **"Shear Plate"** on the Front plane. The lower left corner must be located at the Front plane's origin. Sketch the general shape first, then add the dimensions and sketch relations in alphabetical order around the sketch. **Design Intent** – The left hole must be directly above the lower left corner "A" at a height of 1.00 inches with a diameter of 0.50 inches. The right hole must be the same size and along a horizontal axis 2.00 inches to the right. Line lengths are specified in the figure along with specified angles between lines. Note that line EF is perpendicular to line DE. Line GH is also perpendicular to line EF. Line FG is parallel to line BC. Lines AB and HJ are horizontal. After adding these dimensions and sketch relations, add a Driven dimension for line AJ because its length is predetermined. What is the length of line AJ? _____ inches

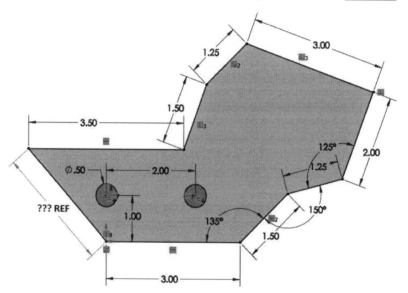

2-2   Draw the **"Rocker"** on the Front plane with the origin at the center of the rocker. Be sure to draw the horizontal and vertical centerlines. All arcs are tangent to one another.

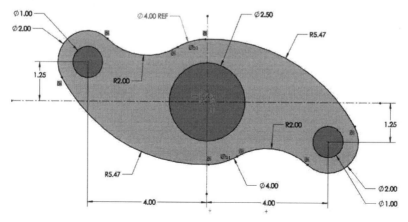

2-3   Draw the **"Latch Plate"** on the Front plane with the origin at the midpoint on the left edge. Note that this part is symmetrical so be sure to add a horizontal centerline.

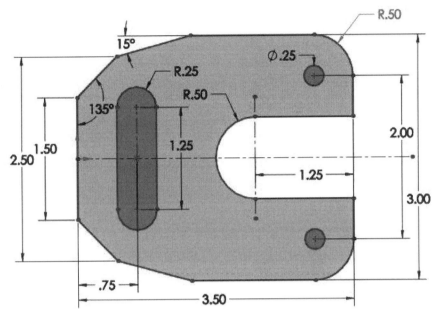

2-4   Draw the **"Special Cam"** on the Front plane. The outer shape is an ellipse centered at the origin. Draw horizontal and vertical centerlines through the origin as shown.

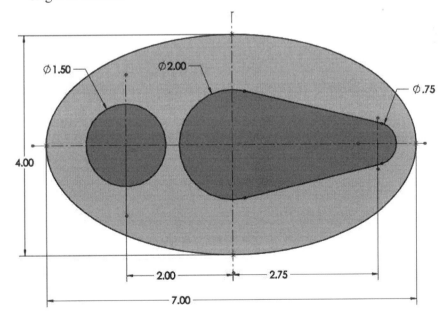

2-5   Use the center rectangle tool to draw the **"Cover Plate"** on the Front plane with the origin at the center of the cover plate. Use the linear pattern tool to create the 4-hole pattern. Since the part is symmetrical, Draw horizontal and vertical centerlines through the origin. (Not shown here.)

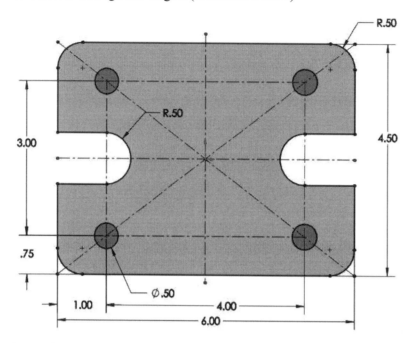

2-6   Draw the **"Front Cover Plate"** on the Front plane with the origin in the lower-left corner of the plate. All dimensions are in millimeters. Be sure to change your drawing units from inches to millimeters in the lower right corner of the window.

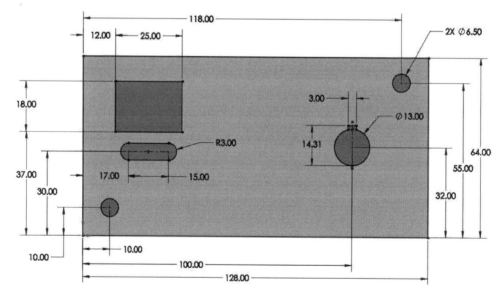

2-7   Draw the **"Support Frame"** on the Front plane with the origin at the lower left corner of the support frame. After adding all necessary dimensions and sketch relations add three driven dimensions so you can determine the two vertical line lengths and the horizontal span of the support frame. If the three REF dimensions in the sketch below are not driven dimensions, then you are missing a dimension or a sketch relation.

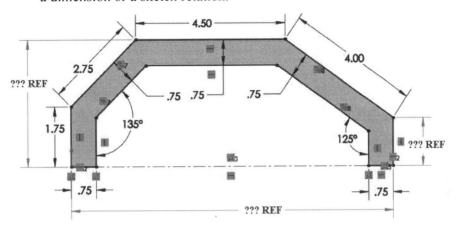

2-8   Draw the **"Nylon Gasket"** on the Front plane with the origin at the center of the gasket. Initially draw three centerlines from the origin, spaced 120 degrees apart. All the radial arcs are tangent to each other. Similar size arcs are equal in size. Use the Equal sketch relation so you don't have to dimension each arc. Hint: Draw circles at appropriate locations, then use the Trim Entities tool to remove unwanted sections of the circles.

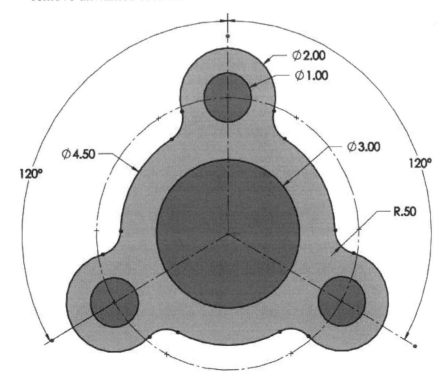

2-9   Draw the **"Breather Plate"** with the origin in the lower-left corner of the plate.
        Use the linear pattern tool to create the 12-hole pattern. Use the **Equal** sketch
        relation to make sure all four rounded corners are the same size at 0.50 inches.

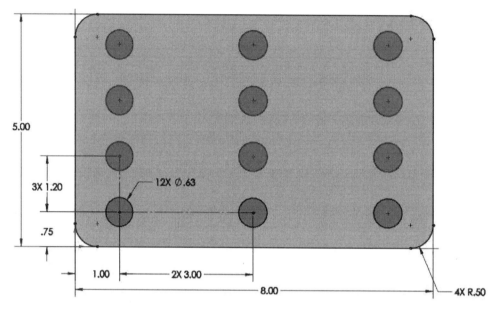

2-10   Draw the **"Three-Quarter Moon Disk"**. Draw horizontal and vertical center-
        lines through the origin as shown.

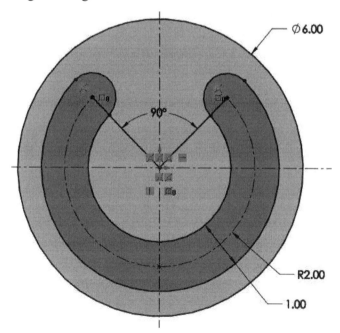

Chapter 3.   Extruded Boss/Base

Introduction

At this point, you should be familiar with sketching shapes. If not, go back and work through the previous chapter because sketching is the main tool of part creation. This section introduces extrusions, which are 2D area sketches extended into a third dimension, thus creating a three-dimensional volume. These extrusions can be positive, which adds material, or negative, which subtracts material. The first extrusion is always positive and is referred to as the parent. The additional extrusions will be child features of the first extrusion or previously defined extrusions.

A child feature is dependent upon the parent feature. To create a block with a hole, first, create the block (parent), and then add the hole (child). You cannot create the hole, and then add the block because the child cannot come before the parent.

Before we begin creating extrusions in SOLIDWORKS let's practice visualizing how to create a part by drawing one of its cross-sections and extending it in the third dimension to create a volume. For example, let's visualize the creation of a block with two holes as shown.

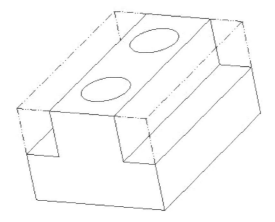

We could create the upside-down tee-shaped cross-section, extend it into the third dimension, and then add the two holes.

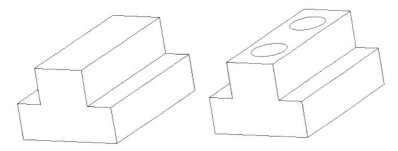

We could create the wide bottom section, and then add the top section and finally the two holes.

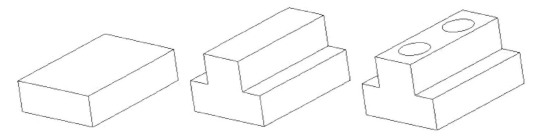

We could create the middle section with two holes and then add the lower extensions.

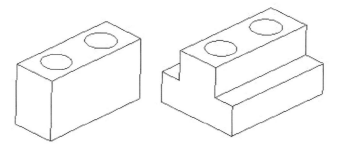

As you can see there are several ways to create this part. Based on design intent, one way may be better than the others, but each way did create the same part. Don't be too concerned if you are creating a part in a different order than your classmates. The important thing is: "Can you create the part?"

Let's look at block 15703 which can be created from extrusions. How would you create this part?

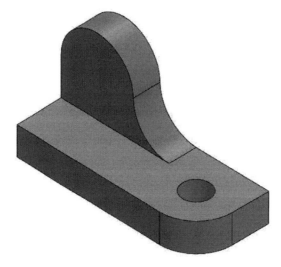

If we assume the Top plane in SOLIDWORKS is the bottom of the part, we would sketch the footprint of the part on the Top plane. Remember that the sketched section below must be a closed section to extrude it.

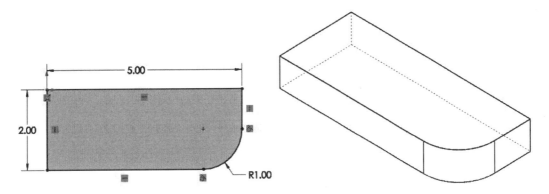

Assuming the Front plane is the back of this part, sketch the protrusion above the base, and then extrude it forward the specified thickness.

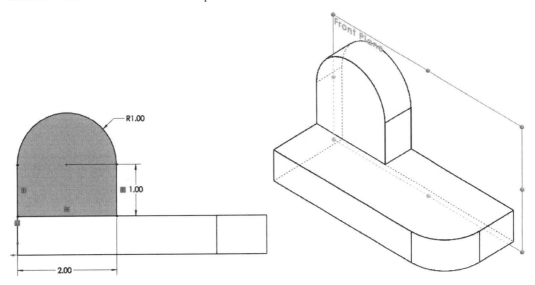

Next, we can add the round or fillet where the upper protrusion meets the base protrusion, and finally add the hole in the base section.

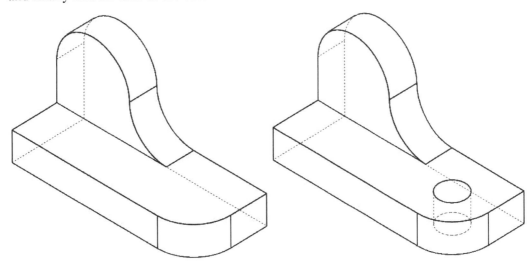

Let's look at an adjustable sliding axle support. How would you create this part? The paper scale is laid out in inches.

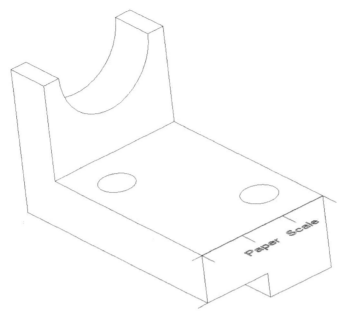

Assuming the paper scale is on the right end of the part, sketch its cross-section, and then extrude it toward the left the entire length of the part.

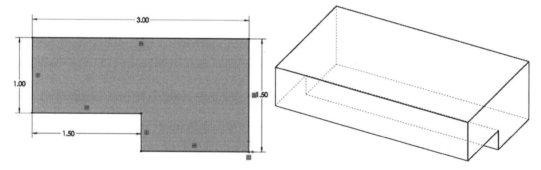

Use the left end of the part as the sketching plane, sketch the protrusion above the base section, and then extrude it to the appropriate width. Note that the top edge of the cutout of the part is 0.38 inches on each side.

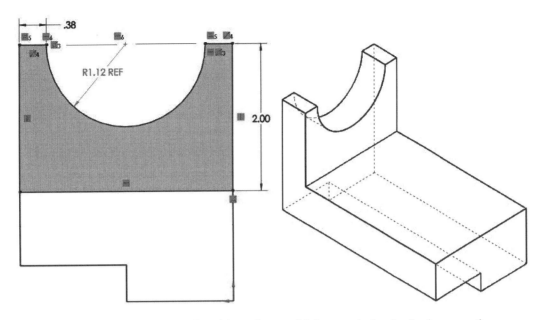

Using the top of the base as a sketching plane, add the two holes in the base section.

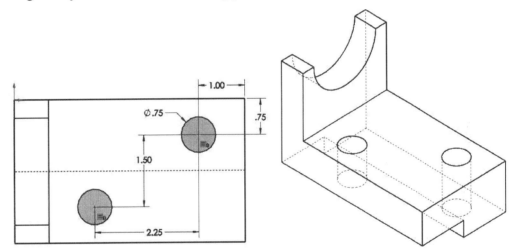

Let's look at a support bracket. Its construction should be obvious. (The cutaway is used to show the second hole.)

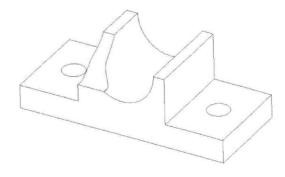

Sketch this view on the Front plane, then extrude it to the appropriate depth. Let's assume the origin is in the exact middle of the bottom plane of the part. Add relation constraints to make the part symmetric.

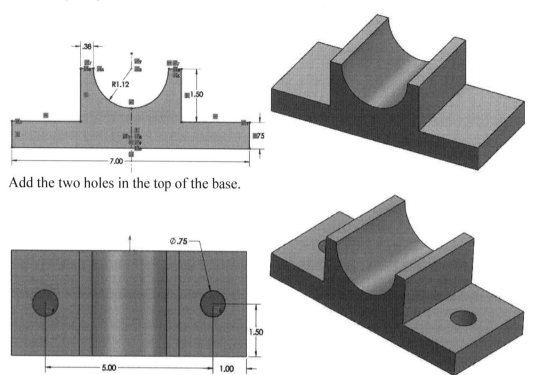

Add the two holes in the top of the base.

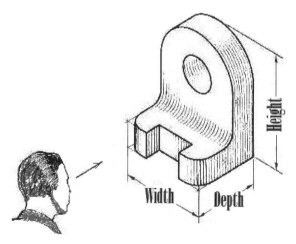

The man pictured below is looking at the front of this part. The origin of this part is directly below the hole on the back side of the part. How might you create this part? Are you seeing a pattern yet?

One approach might be to sketch the footprint of the bracket which will define the width and depth of the part along with the front cutout and rounded front corners. Note that the origin is located in the middle of the back edge of the part.

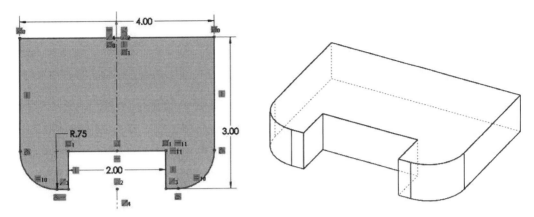

Add the upper protrusion by sketching on the Front plane and extruding it forward to the appropriate thickness.

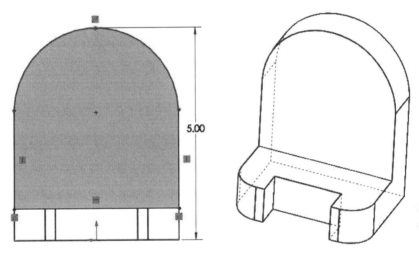

Add the hole to the top protrusion centered at the radius center, and finally add a rounded edge between the two protrusions. Note that these two operations could be done in either order.

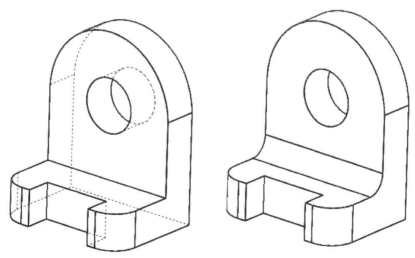

Do the constructions make sense?

Describe how you might construct the following part in SOLIDWORKS. The lower rectangular slot goes all the way through the part. The 1.00-inch diameter hole goes through the upper portion of the part.

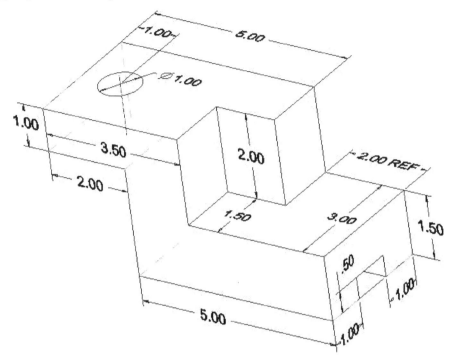

Extruded Boss/Base Practice

New part creation follows a standard procedure in SOLIDWORKS. This procedure is used for most thin-walled and solid extrusions. This text will make several assumptions when you are creating a new part. We will use the **Part_IPS_ANSI** SOLIDWORKS template file when creating a new part. New parts will be assigned a material property, so if any analysis is done requiring the material property, a material will already be defined.

There are several ways to extrude a feature. Let's begin by starting SOLIDWORKS. Be sure to toggle into **Advanced** mode.   Select the **Part_IPS_ANSI** SOLIDWORKS Template, then pick OK.

Design the symmetrical **"Holder Block"** shown below with the origin located in the middle of the lower left edge. The 1.50-inch diameter through-hole is horizontally centered on the slanted plane; thus a symmetrical extrusion is required about the Front plane when creating this part. This part is made from malleable cast iron.

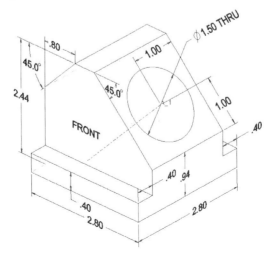

1. If the Sketch and Features tools are not on the sides of the window, you can also add the Sketch and Features tools to the sides of the window. The Sketch tools will appear along the right side of the window. The Features tool will appear along the left side of the window. These are added by using the **Toolbars** option from the same pop-up menu.

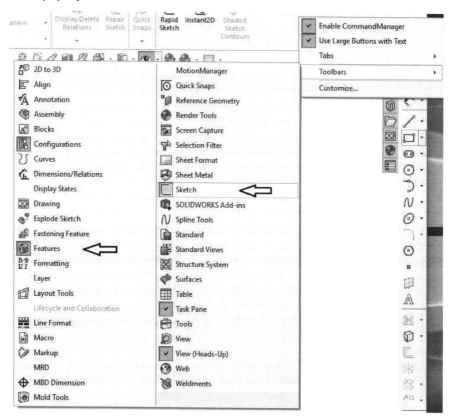

2. Selecting the **CommandManager** from the pop-up menu disables it. Picking it again enables it. Disabling the **Use Large Buttons with Text** option gives you more room to work.

3. Pick the Top plane from the Feature Manager Design Tree, then select the sketch tool.

4. Draw a horizontal centerline through the origin.

5. Draw a rectangle symmetric about the horizontal centerline. Be sure to make the rectangle symmetric about the centerline by adding the symmetric sketch relation. Pick a corner of the rectangle with the LMB, hold down the <Ctrl> key, and pick the corner on the opposite side of the centerline with the LMB, then with the <Ctrl> key still held down, pick the centerline with the LMB. In the Add Relations window pick Symmetric

6. Use the Smart Dimension tool to size the rectangle 2.80 x 2.80 inches.

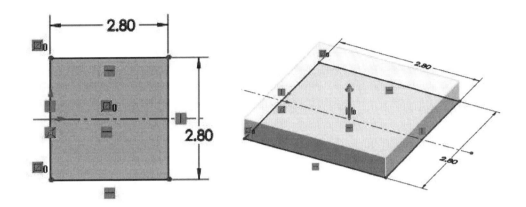

7. Exit Sketch by picking the Exit Sketch icon.

8. With **Sketch1** highlighted in the **Feature Manager Design Tree**, pick the **Extruded Boss/Base** icon.

9. Set the extruded thickness to 0.40 inches. Pick the green checkmark.

10. Select the Front plane from the **Feature Manager Design Tree**, then select the sketch tool.

11. Use the line tool to sketch the general shape of the top portion of the part.

12. Use the Smart Dimension tool to properly side the shape. Note that because of design intent, only the dimensions shown in the figure are permitted on this sketch. The 2.44-inch dimension is needed for clearance, and the 0.94-inch dimension is needed to ensure the part will slide under the protruding lip of the mating part.

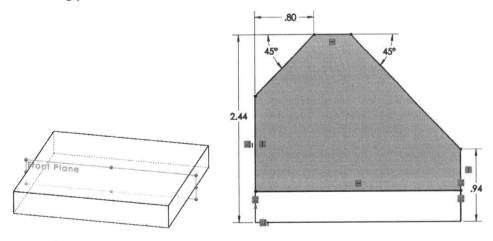

13. Exit Sketch by picking the Exit Sketch icon.

14. ![icon] With **Sketch2** highlighted in the **Feature Manager Design Tree**, pick the **Extruded Boss/Base** icon.

15. For Direction 1, select **Mid Plane** from the pull-down menu. Set the extruded thickness to 2.00 inches. Pick the green checkmark.

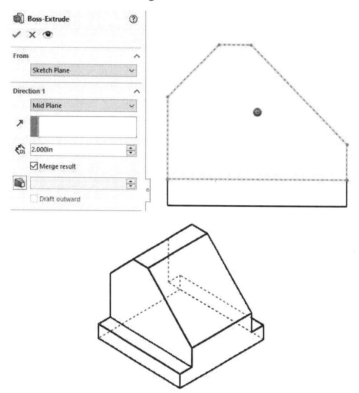

16. ![icon] Select the front inclined face with the LMB, then pick the Sketch icon.

17. ![icon] Draw a circle on this surface, then locate it and size it as shown using the Smart Dimension tool.

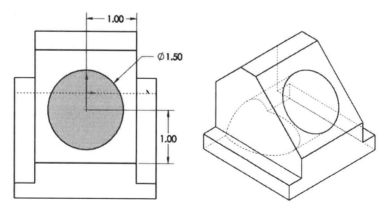

18. 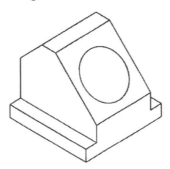 With **Sketch3** highlighted, pick the **Extruded Cut** icon. Set direction 1 to **Through All**, then pick the green checkmark.

19. In the **Feature Manager Design Tree**, pick Material <not specified>, then press down the RMB and hold until a pop-down menu appears. Select **Malleable Cast Iron** from the Material list.

20. **File> Save**. Enter the name **"Holder Block"**. Make sure you are saving the part file in the **SOLIDWORKS Parts** folder. Pick the **Save** button.

21. **File> Print…** (For your records.)

22. **File> Close**

23. Exit SOLIDWORKS if you are done for now, otherwise move on to the next section.

Extruded Boss/Base Exercise

Let's create the **"Stocktail Clamp"** shown below with some modifications.

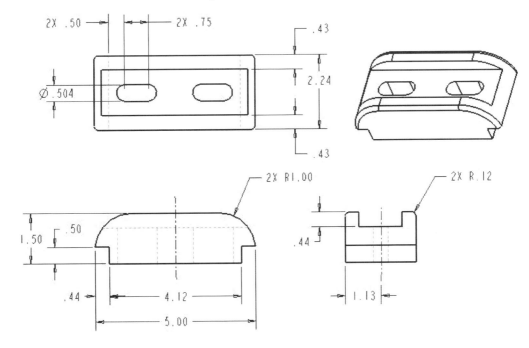

Design Intent – Some dimensions above do not reflect design intent. For example, the 0.43-inch wall thickness at the top of the clamp is not important. Instead, the flat cutout needs to be 1.38 inches wide. The 0.44-inch depth of the top cutout is not important. Instead, the thickness of the center section of the clamp needs to be 1.06 inches. The two slots need to be located along the geometric center with the origin of the clamp on the bottom surface at its geometric center. This part is made from plain carbon steel.

1. If you exited from SOLIDWORKS at the end of the last session, then restart SOLIDWORKS. Be sure to toggle into **Advanced** mode. Select the **Part_IPS_ANSI** SOLIDWORKS Template, then pick **OK**.

2. If the Sketch and Features tools are not on the sides of the window, you can also add the Sketch and Features tools to the sides of the window. The Sketch tools will appear along the right side of the window. The Features tool will appear along the left side of the window. These are added by using the **Toolbars** option from the same pop-up menu.

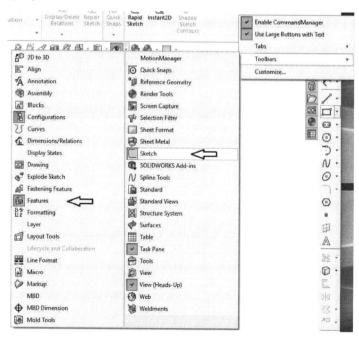

3. Selecting the **CommandManager** from the pop-up menu disables it. Picking it again enables it. Disabling the **Use Large Buttons with Text** option gives you more room to work.

4. Pick the Front plane from the **Feature Manager Design Tree**, then select the sketch tool.

5. Draw a vertical centerline through the origin.

6. Draw a rectangle symmetric about the vertical centerline. Be sure to make the rectangle symmetric about the centerline by adding the symmetric sketch relation. Pick a corner of the rectangle with the LMB, hold down the <Ctrl> key, and pick the corner on the opposite side of the centerline with the LMB, then with the <Ctrl> key still held down, pick the centerline with the LMB. In the Add Relations window pick Symmetric.

7. Use the Line tool to create two cutouts in the lower left and right corners.

8. Use the Trim Entities tool to remove any unwanted lines.

9. Use the Smart Dimension tool to size the rectangle 1.50 x 5.00 inches. Use the Smart Dimension tool to properly side the two cutouts as shown.

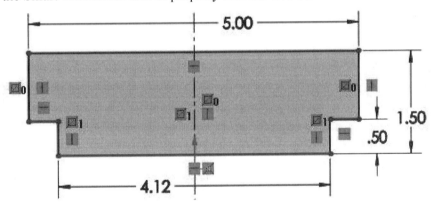

10. Exit Sketch by picking the Exit Sketch icon.

11. With **Sketch1** highlighted in the **Feature Manager Design Tree**, pick the **Extruded Boss/Base** icon.

12. For Direction 1, select **Mid Plane** from the pull-down menu. Set the extruded thickness to 2.24 inches. Pick the green checkmark.

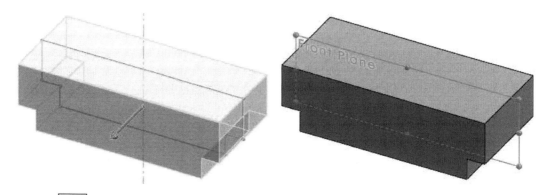

13. 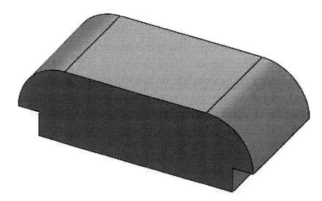 Use the Fillet tool on the 3D part to round the two upper corners using a 1.00-inch radius fillet as shown.

It is always a good idea to save your parts regularly especially as they become more complex.

14. **File> Save**. Enter the name **"Stocktail Clamp"**. Make sure you are saving the part file in the **SOLIDWORKS Parts** folder. Pick the **Save** button.

15. In the View Heads Up toolbar, under Display style, pick the **Hidden Lines Removed** icon to show a stick figure with no hidden lines.

16. Pick the Front plane, then pick the Sketch icon again.

17. Use the Line tool and the Smart Dimension tool to draw a horizontal line 1.06 inches from the bottom of the part, then add three more lines to make a closed area as shown. The shaded area can be above the part without any issues.

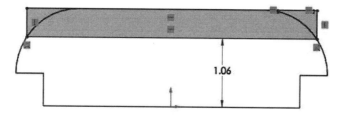

1.06

18. Exit the Sketch.

19. With **Sketch2** highlighted in the **Feature Manager Design Tree**, pick the **Extruded Cut** icon.

20. Set Direction 1 to **Mid Plane**, and the width of the cut to 1.38 inches. Pick the green checkmark.

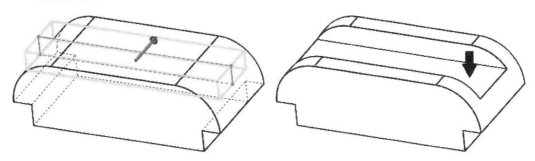

21. Select the cutout surface with the LMB, then select the Sketch icon.

22. Draw a vertical centerline through the origin.

23. Use the **Straight Slot** tool to draw two horizontal straight slots.

24. In the View Heads Up toolbar, select the **Hidden Lines Visible** to show hidden lines.

25. Make the slots symmetric about the vertical centerline and equal in radius.

26. Use the Smart Dimension tool to position and size the slots as shown. For the slot radius dimension enter .252, then change the Precision to .123 in the Dimension window. The faded .75 REF dimension is a driven dimension.

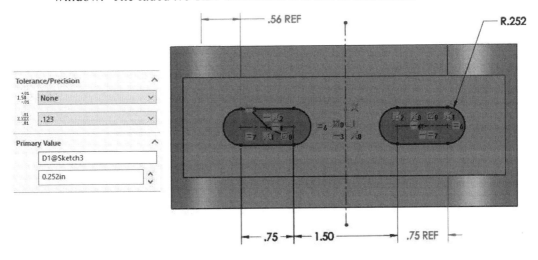

27. Exit Sketch.

28. With **Sketch3** highlighted in the **Feature Manager Design Tree**, pick the **Extruded Cut** icon. Set Direction 1 to **Through All**. Pick the green checkmark.

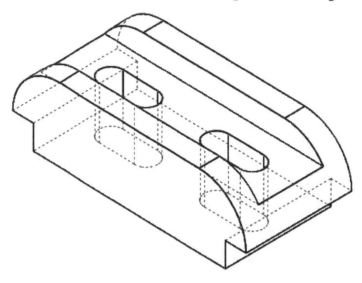

29. Use the Fillet tool on the 3D part to round the outer edges using a 0.12-inch radius.

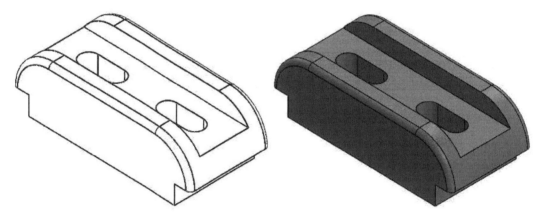

30. In the **Feature Manager Design Tree**, pick Material <not specified>, then press down the RMB and hold until a pop-down menu appears. Select **Plain Carbon Steel** from the Material list.

31. **Tools>Evaluate>Mass Properties…** will allow you to determine the mass and volume of this part. The mass should be 3.12 pounds and the volume should be 11.06 in^3. If your values are different, go back and check your work.

32. **File> Save**.
33. **File> Print…** (For your records.)
34. Pick **Close**

35. Exit SOLIDWORKS if you are done for now, otherwise move on to the next section.

Extruded Boss/Base Problems

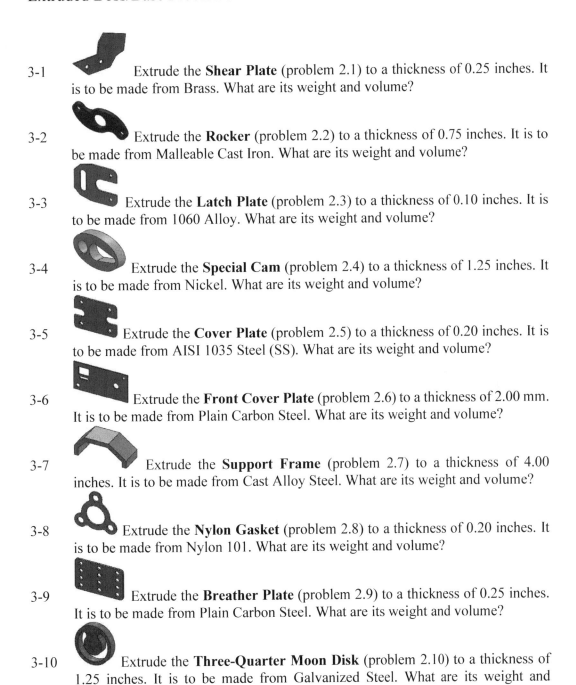

3-1   Extrude the **Shear Plate** (problem 2.1) to a thickness of 0.25 inches. It is to be made from Brass. What are its weight and volume?

3-2   Extrude the **Rocker** (problem 2.2) to a thickness of 0.75 inches. It is to be made from Malleable Cast Iron. What are its weight and volume?

3-3   Extrude the **Latch Plate** (problem 2.3) to a thickness of 0.10 inches. It is to be made from 1060 Alloy. What are its weight and volume?

3-4   Extrude the **Special Cam** (problem 2.4) to a thickness of 1.25 inches. It is to be made from Nickel. What are its weight and volume?

3-5   Extrude the **Cover Plate** (problem 2.5) to a thickness of 0.20 inches. It is to be made from AISI 1035 Steel (SS). What are its weight and volume?

3-6   Extrude the **Front Cover Plate** (problem 2.6) to a thickness of 2.00 mm. It is to be made from Plain Carbon Steel. What are its weight and volume?

3-7   Extrude the **Support Frame** (problem 2.7) to a thickness of 4.00 inches. It is to be made from Cast Alloy Steel. What are its weight and volume?

3-8   Extrude the **Nylon Gasket** (problem 2.8) to a thickness of 0.20 inches. It is to be made from Nylon 101. What are its weight and volume?

3-9   Extrude the **Breather Plate** (problem 2.9) to a thickness of 0.25 inches. It is to be made from Plain Carbon Steel. What are its weight and volume?

3-10   Extrude the **Three-Quarter Moon Disk** (problem 2.10) to a thickness of 1.25 inches. It is to be made from Galvanized Steel. What are its weight and volume?

3-11 Design the symmetrical "**Angle Brace**" with the origin located in the middle of the lower left edge. The 0.75-inch diameter through-hole is horizontally centered on the slanted plane; thus a symmetrical extrusion is required about the Front plane when creating this part. All dimensions are in inches. The part is made from Plain Carbon Steel. Remember to change to three decimal places for the .375 diameter dimension.

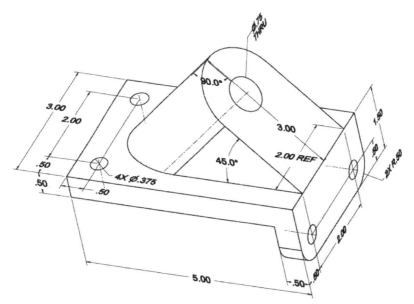

3-12 Design the symmetrical "**Hanging Bracket**" made from Malleable Cast Iron. The 1.75-inch diameter hole is located 9.00 inches from the lower left edge. The straight portion of the rightmost extension starts 2.50 inches from the lower hole's center and is 2.25 inches long. **Design change:** the two 2.00-inch diameter holes need to be increased to 2.03 inches. All dimensions are in inches.

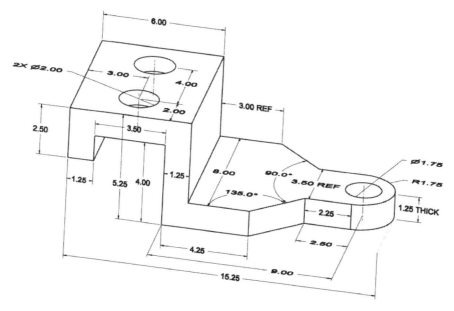

3-13   Design the "**U-Brace**" with the origin located at the center of the lower 0.50-inch diameter hole. The center of the upper arc is 2.25 inches above the bottom of the part. The right end is 0.75 inches above the bottom of the part. The part is made from 1060 Alloy. All dimensions are in inches. What is the weight of this part?

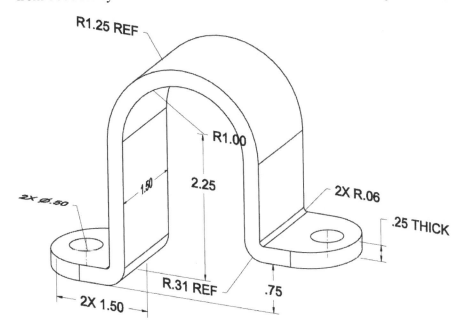

3-14   Design the symmetrical "**Bearing Plate**" made from Plain Carbon Steel. Place the origin at the bottom of the center hole. The two slots must be located 3.25 inches from the center hole located at the origin. **Design change:** increase the center .75-inch diameter hole to .88 inches. All dimensions are in inches.

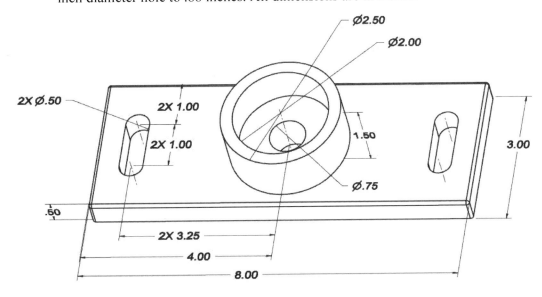

3-15 Design the **"Slotted Cam"** made from Cast Alloy Steel with the origin located at the center of the 1.50-inch diameter through-hole. The 3/4-inch diameter slot should follow a path similar to the outside top edge of the cam. Use an extruded cut to create the curved slot. Place a datum point 4.75 inches from the origin before making the curved slot. Note that some dimensions require three decimal places as shown. All dimensions are in inches.

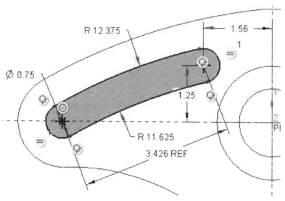

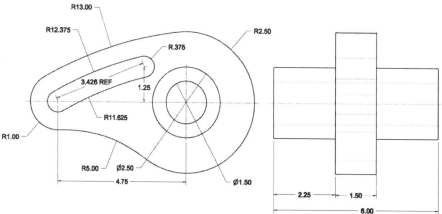

3-16 Design the symmetrical **"Operating Arm"** with the origin located at the back of the 1.75-inch diameter hole. This part is made from Plain Carbon Steel. All dimensions are in inches. The R4.00 radius is centered at the origin. The end of the 3-point arc slot at 30° and 15° from the vertical centerline.

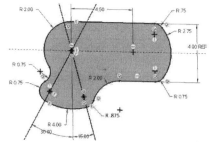

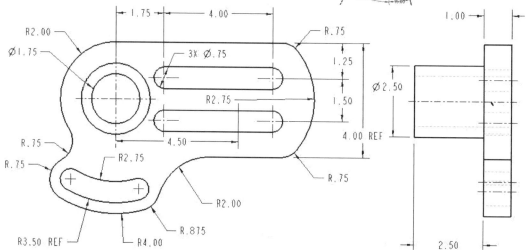

3-17   Design the symmetrical "**Cutoff Holder**" made from Malleable Cast Iron with the origin located in the middle of the bottom leftmost edge. The vertical hole must be 1.50 inches from the right end and 1.06 inches from the front surface. The horizontal hole must be 2.094 inches above the top of the flat cutout (not the bottom of the part) and in line with the vertical hole. All dimensions are in inches. Remember to change to three decimal places when the dimensions are three decimal places in the figure below.

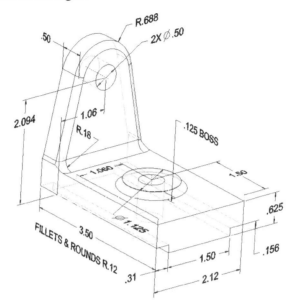

3-18   Design the symmetrical "**Bearing Holder**" made from Plain Carbon Steel with the origin located in the back center of the part. The ¼-inch drill goes through the top wall of the part only. Add the 1/8-inch fillets and rounds last. **Design change**: The design engineer wants a .03x45° chamfer (not shown) added to both ends of the 1.625-inch diameter hole. All dimensions are in inches. Remember to change to three decimal places when shown in the figure below.

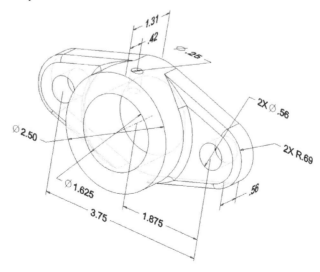

3-19 Design the **"Clutch Lever"** made from Plain Carbon Steel with the origin located at the front center of the large 1.250-inch diameter reamed, through hole. The top 3/8-inch hole is centered between the front and back planes of the part and goes all the way through the part. The counter bore is located at the bottom of the part. The two 3/8-inch diameter holes go through their respective walls.

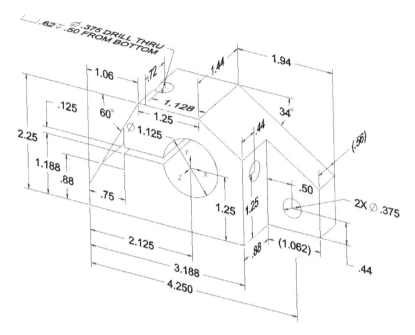

3-20 Design the symmetrical **"Boiler Stay"** made from Plain Carbon Steel. Place the origin at the bottom of the leftmost hole. The three through holes must be located on 50-millimeter centers. **Design change:** The 20-millimeter diameter through-hole must be 106 mm from the leftmost hole's center. Also, decrease the three 13-mm diameter holes to 12.5 mm. All dimensions are in millimeters.

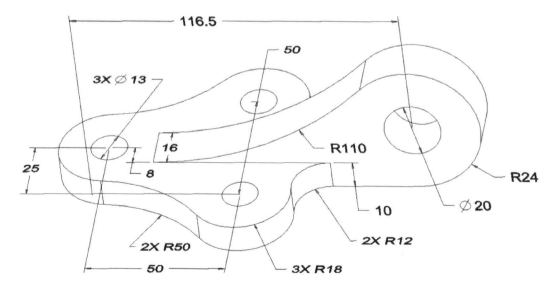

3-21   Design "**Vibrator Arm**" made from Malleable Cast Iron. The bottom slot goes through the bottom portion of the part. The two vertical through-holes must be 1.88 inches apart. The 0.593 dimension becomes a reference dimension. The preferred origin is at the back of the left 0.750-inch diameter hole. The right, lower flat section is 1.50 inches long. All dimensions are in inches. **Design change:** decrease the two 0.750-inch diameter holes to 0.725 inches.

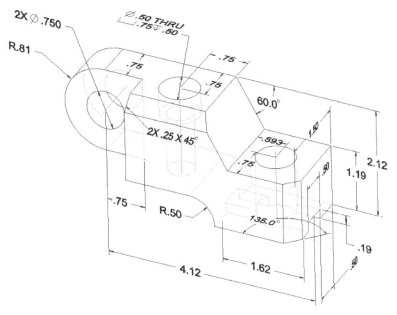

3-22   Design the "**Operating Lever**" made from Malleable Cast Iron with the origin at the bottom of the large 1.250-inch diameter through hole. The keyway slot is ¼ inch wide and goes through the entire part. The distance from the top of the keyway to the opposite side of the hole is 1.365 inches. The center of the slot aligns with the center of the hole. **Design change:** The slot needs to be 0.762 inches wide instead of 0.75 inches wide. What is the weight of this Operating Lever?

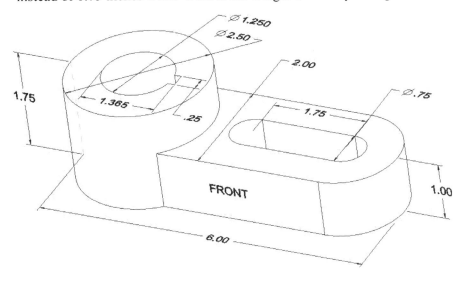

3-23   Design the **"Slide Block"** made from Cast Alloy Steel. Design change: The ½-inch hole needs to include the ream operation as part of the dimension.

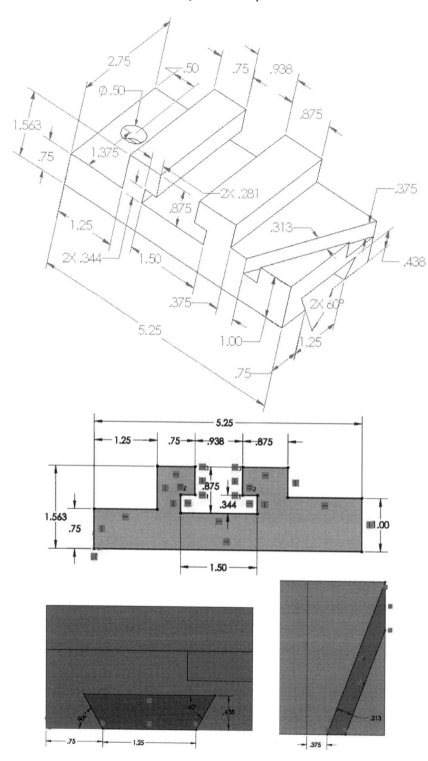

3-24　Design the **"Clamp Base"** made from plain carbon steel. The front two corners use a .25-inch radius. Use a .06-inch round of the top edge and the bottom edge as shown in the figure. **Design change:** create the 25/64" holes at ϕ.392 inches.

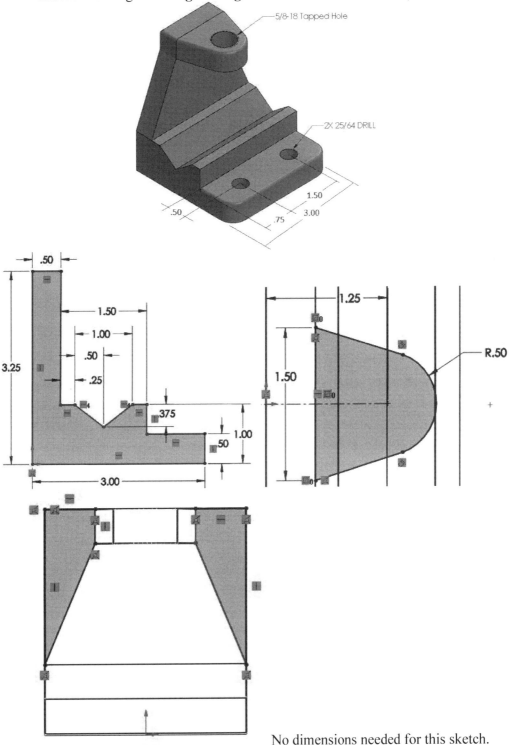

No dimensions needed for this sketch.

3-25   Design the **"Feeder Bracket"** made from plain carbon steel. Add a general note stating that All Fillets and Rounds are .06-inch radii.

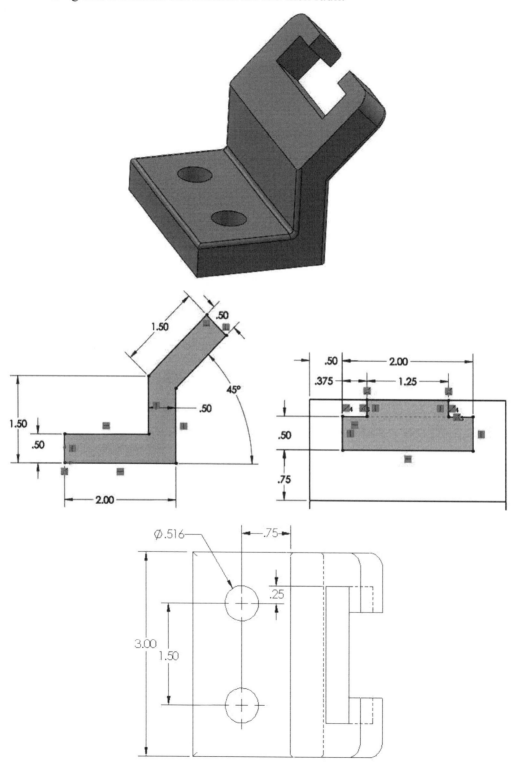

3-26   Design the **"Guide Bracket_430"** made from plain carbon steel. Add a general note stating that all fillets and rounds have a .06-inch Radius unless otherwise specified. Use a 25/32 DRILL (ϕ.781) to create the two holes in the base of the part.

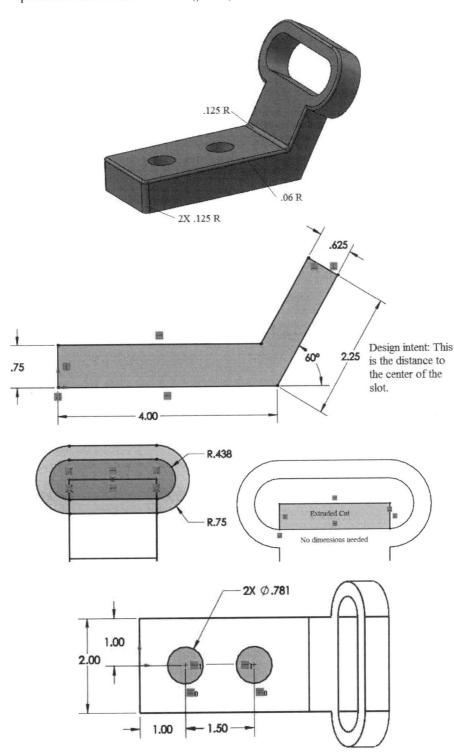

3-27 Design the **"Guide"** made from cast alloy steel. Be sure to use the dimensions shown in these sketches when creating this part. You will be adding Geometric Dimensions and Tolerances to this Guide in Problem 10-7. Add a general note stating that All Fillets and Rounds are .06-inch radius.

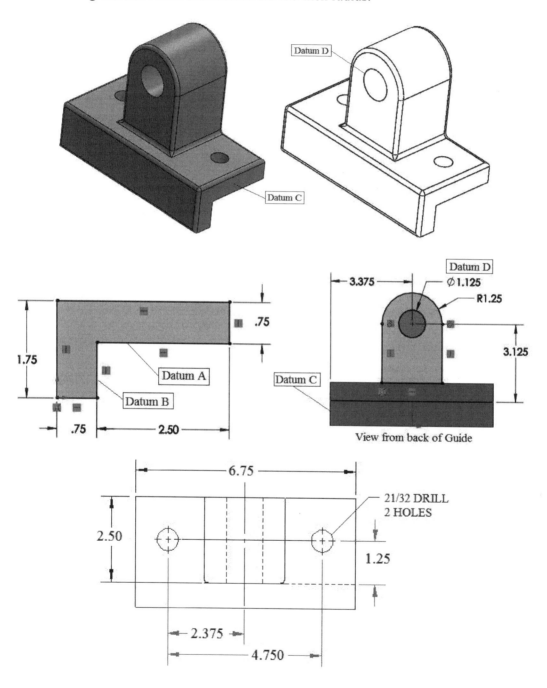

3-28   Design the **"Offset Bracket"** made from plain carbon steel. Add a general note stating that All Fillets and Rounds are .06-inch radius.

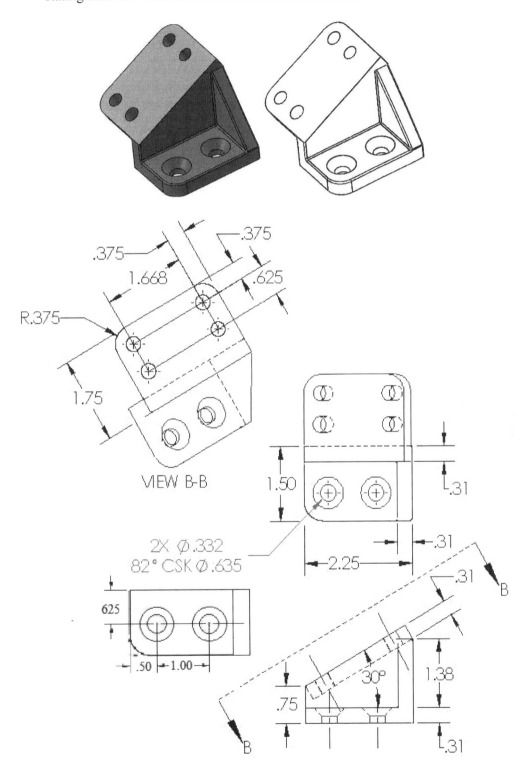

.375

.375

1.668

.625

R.375

1.75

VIEW B-B

1.50

.31

2X Ø.332
82° CSK Ø.635

2.25

.31

625

.50   1.00

.31   B

.31

30°

1.38

.75

.31

B

3-29   Design the **"Guide Bracket_302"** made from cast alloy steel. **Design change:** add a .25-inch radius to the inner corner of the part to reduce the stress concentration factor. Also, add a .25-inch fillet where the rounded portion meets the right angle portion of the part.

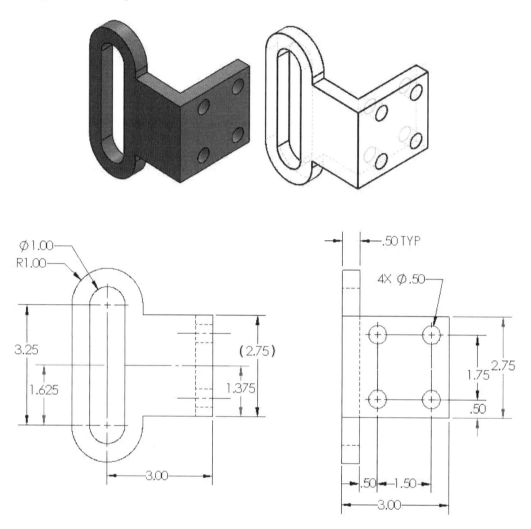

3-30   **Challenge problem** - Create the **"Torch Holder"** shown below. The part is made from plain carbon steel. Convert 1/8th inch and 1/16th inch fractional dimensions to 3-place decimals. Use 2-decimal places for all other fractional dimensions. **Design change:** the 5/16-18 NC-2 threads need to be changed to 3/8-16 UNC-2. The 11/32-inch drilled clearance hole needs to be increased to a "W" (.386) drill clearance hole. (It may be easier to create this part after finishing Chapter 4.)

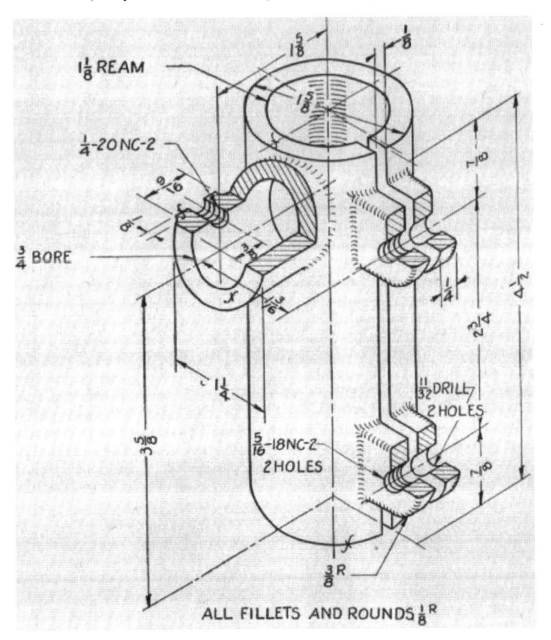

ALL FILLETS AND ROUNDS $\frac{1}{8}$R

3-31   **Challenge problem** - Create the "**Tension Bracket**" shown below. The part is symmetrical about its centerline. The part is made from 1020 CD steel. Convert all fractional dimensions to 3-place decimals. After testing the design engineer noted that a small crack had formed in the sharp underneath corner so he/she decided to place a 1/4-inch round in this corner (not shown). Modify your part accordingly. **Design change:** After further testing, the engineer has decided to increase the horizontal plate thickness from 7/16-inch to 1/2-inch. Also, add a general note to break all sharp edges.

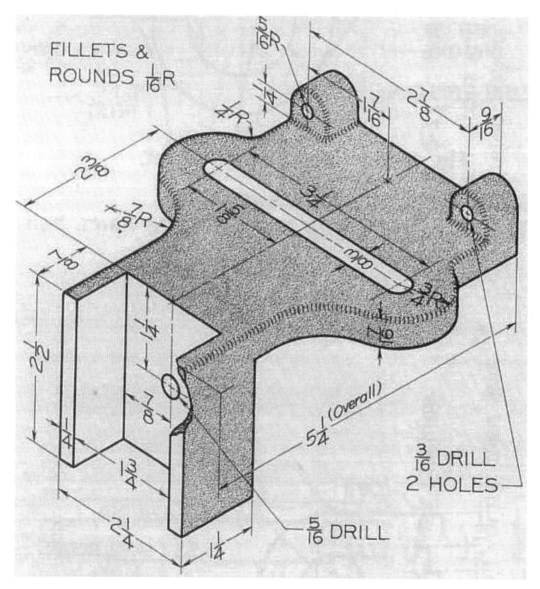

Chapter 4.   Revolved Boss/Base

Introduction

Symmetrical Features in Designs

Symmetry is an important characteristic seen in many designs. Symmetrical features can be created by an assortment of tools available in SOLIDWORKS. You can create multiple copies of a symmetrical feature with the Feature Pattern command, or you can create a mirrored image of the model or a 2D section using the Mirror command.

It is important to identify the features that exist in a design when doing parametric modeling. Feature-based parametric modeling enables you to build complex designs by creating a series of simple features. This approach simplifies the modeling process and allows you to concentrate on the features of the design.

The modeling technique of extruding two-dimensional sketches perpendicular to the sketch to create three-dimensional features was discussed in the previous chapter. For cylindrical parts with tapered surfaces, the protrusion procedure will not work. For designs that involve cylindrical shapes revolving a 2D sketch about an axis will form the needed 3D feature. In solid modeling, this is called a revolved feature.

The small tapered clutch was created using the Revolved Boss/Base command in SOLIDWORKS. Because the clutch has a taper, it cannot be created using the Extrude Boss/Base command. The keyway can be created using an Extruded Cut through the clutch.

A revolve can be created by sketching the cross-section in the sketch, exiting the sketch, highlighting it in the **Feature Manager Design Tree**, and then picking the Revolve Boss/Base tool. For this example, we are going to sketch the upper cross-section of the clutch along with its centerline, then revolve the sketch. Since the section will be revolved, the diameter dimensions should be used. **Adding a diameter dimension is done by picking the centerline and the diameter point, then moving the mouse cursor below the centerline and pressing the LMB to place the diameter dimension at the current cursor position.** When you have finished dimensioning the section, exit sketch, and revolve the section.

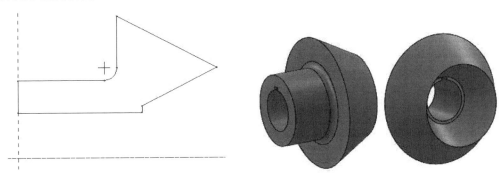

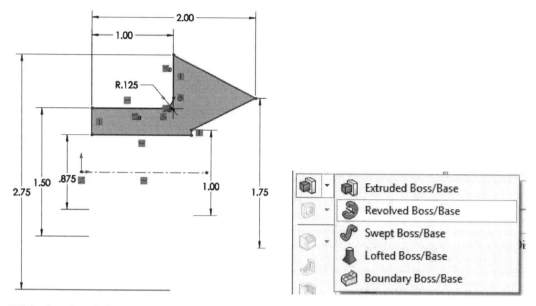

With the sketch highlighted, pick the Revolve Boss/Base icon from the left side of the screen. The Revolve tab will appear. You will use the centerline you drew in the sketch as the axis of rotation. Instead of depth when extruding a feature, revolves have an angle of rotation for the sketched section. A complete 360-degree rotation of this section is necessary to create the small tapered clutch.

Use the mouse to view the part from different locations. When all appears correct, select the green checkmark to accept the revolved feature.

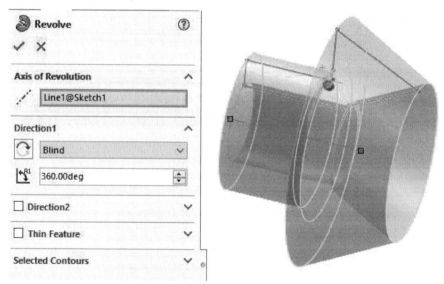

If you need to edit the revolved section, highlight Revolve 1 in the **Feature Manager Design Tree**, then press and hold down the LMB until a pop-up menu appears. Pick the Edit Feature icon from this menu to reactivate the Revolve window.

If you need to change the sketch or its dimensions, you can edit the sketch by picking the arrow sign in front of Revolve 1 to get Sketch 1 to show up. With Section 1 highlighted, press down and hold the LMB until a pop-up menu appears. Select the Edit Sketch icon from this pop-up menu to get back into sketch mode. When finished, exit the sketch.

The keyway is added using an extrude cut with the left end of the clutch selected as the sketching plane. The hole in the center of the clutch is used as a reference for dimensioning. The Corner Rectangle tool is used to create a rectangle that is 0.125 inches and coincident with the edge of the hole. To show design intent, the width of the keyway and the distance across the hole to the top of the keyway are dimensioned. <u>First, place a Point on the opposite side of the hole, then use the Smart Dimension tool to create a dimension between the newly added point and the top of the rectangle.</u> The machinist will use these two dimensions when cutting the keyway.

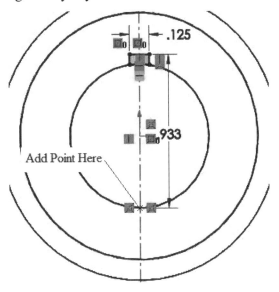

When finished, exit sketch. With Sketch 2 highlighted, select the Extruded Cut icon. Select Through All. Make sure the arrow points into the part. If all looks correct, select the green checkmark to accept the feature. The small tapered clutch is complete.

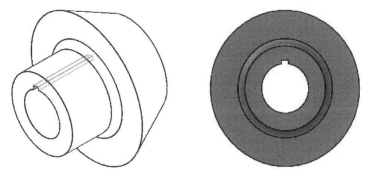

The second example is a 4-hole hub as drawn in CAD with sketched dimensions shown. This basic hub without the four holes, fillets, and rounds can be created using four different extrudes or it can be created using the Revolved Boss/Base tool once.

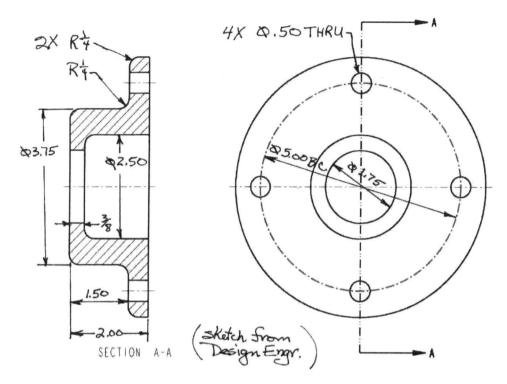

Using the extrude tool the building procedure might look like this. First, draw the large diameter of the hub, and then extrude it 0.50 inches thick. Note that this dimension doesn't represent design intent.

Second, draw the smaller hub portion using the large hub's surface as the sketching plane and extrude it 1.50 inches.

Third, using the RIGHT datum plane as the sketching plane, sketch the large hole on this plane, then do an extruded cut to a depth of 1-5/8 inches. Note that the 1-5/8-inch dimension is not design intent.

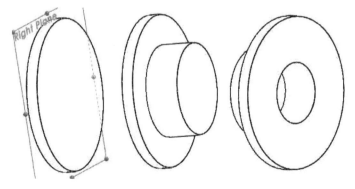

Fourth, using the left surface of the smaller hub as the sketching plane, sketch the small hole on the left side of the hub, then do an extruded cut through the part.

Now all that is left is to round the edges and add the four-hole pattern.

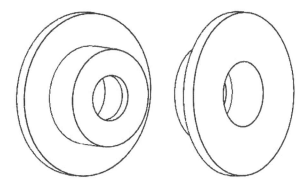

The procedure might look like this if we created it as a revolved feature. First, we would sketch a horizontal centerline and the upper cross-section of the hub, and then revolve it through 360 degrees as shown below. We are done in one step with the sketch representing the engineer's design intent.

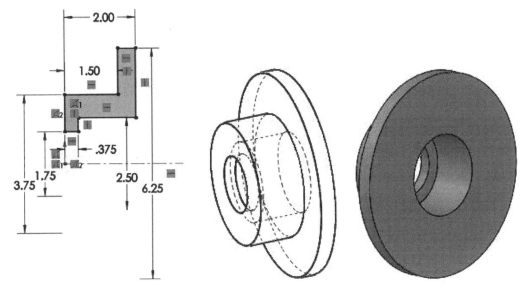

Now all that is left is to round the edges and add the four holes.

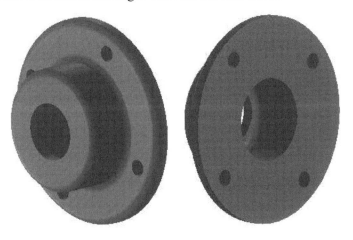

Besides using the Revolved Boss/Base tool, SOLIDWORKS has other tools for handling cylindrical features. The Hole Wizard is one of these.

Cylindrical Options

SOLIDWORKS provides four cylindrical types: the Hole Wizard, Advanced Hole, Thread, and Stud Wizard. For the beginning user, creating a circle, and then doing an Extruded Cut is a much simpler and easier way to create holes. However, the Hole Wizard does work well for countersunk and counter-bored holes up to 1 inch in size for the educational version of SOLIDWORKS.

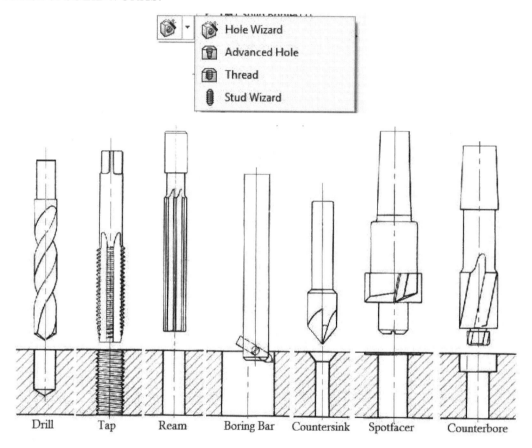

Revolved Boss/Bass Practice

Design a 7.50-inch long **"Step Shaft"** with a 1.00-inch diameter that is 2.00 inches long, a 1.75-inch diameter that is 3.00 inches long, and the remaining length is 1.25 inches in diameter. Both ends of the step shaft need a 0.06-inch x 45° chamfer. The places where the smaller diameters meet the larger diameter need to be rounded using a 0.06-inch radius to reduce stress concentrations. An SH-125 retaining ring needs a minimum groove width of 0.056 inches by 0.037 inches deep placed on the 1.25-inch diameter shaft 2.00 inches from the 1.75-inch diameter. (The machinist needs the inner diameter of the groove which is 1.17 inches and its width to be 0.06 inches. The origin needs to be at the center of the step shaft where the diameter changes from 1.00 inch to 1.75 inches. The shaft is made from AISI 1045 cold-drawn steel.

New part creation follows a standard procedure in SOLIDWORKS. This procedure is used for most thin-walled and solid revolves. This text will make several assumptions when you are creating a new part. We will use the **Part_IPS_ANSI** SOLIDWORKS template file when creating a new part. New parts will be assigned a material property, so if any analysis is done requiring the material property, a material will already be defined.

  1. Let's begin by starting SOLIDWORKS. Be sure to toggle into **Advanced** mode. Select the **Part_IPS_ANSI** SOLIDWORKS Template, then pick OK.

2. If the Sketch and Features tools are not on the sides of the window, you can also add the Sketch and Features tools to the sides of the window. The Sketch tools will appear along the right side of the window. The Features tool will appear along the left side of the window. These are added by using the **Toolbars** option from the same pop-up menu.

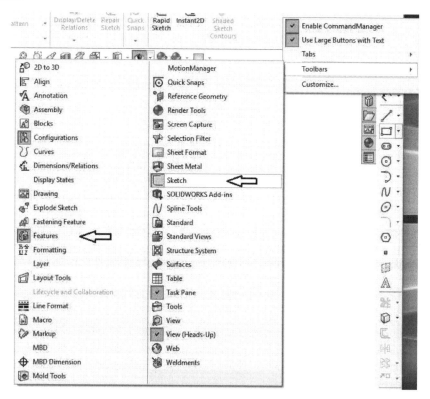

3. Selecting the **CommandManager** from the pop-up menu disables it. Picking it again enables it. Disabling the **Use Large Buttons with Text** option gives you more room to work.

4. Pick the Front plane from the **Feature Manager Design Tree**, then select the sketch tool.

5. Draw a horizontal centerline through the origin.

6. Use the Line tool to create the shaft profile above the horizontal centerline.

7. Pick the centerline then hold down the <Ctrl> key and pick the origin. In the Add Relations window, pick Coincident. Make the solid line on top of the centerline and the origin Coincident as well. Add a vertical sketch relation to make the step change in the shaft diameter directly above the origin.

8. Use the Smart Dimension tool to dimension the sketch.

Adding a diameter dimension is done by picking the centerline and the diameter point with the LMB, and then moving the mouse cursor below the centerline and pressing the LMB to place the diameter dimension at the current cursor position. You can also pick the point, then the centerline.

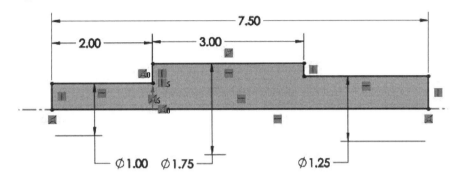

9. Exit Sketch.

10. Pick the Revolved Boss/Base icon to revolve the section about the centerline through 360 degrees. Pick the green checkmark. (Note that the Revolved Boss/Base icon is hidden behind the Extruded Boss/Base icon.)

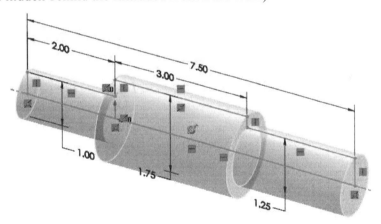

11. Be sure the origin is at the step change in the shaft's diameters. The dot in the figure below represents the origin of this part.

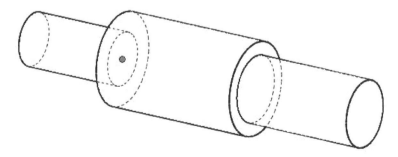

Design intent requires that the ends of the step shaft have a 0.06-inch x 45° chamfer.

12. Use the Chamfer tool to add the four chamfers to the ends of the step shaft. (Note that the chamfer tool is hidden behind the Fillet tool.)

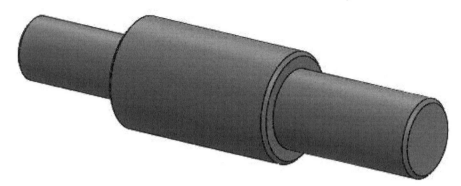

13. Use the Fillet tool to round the two areas where the smaller diameters touch the larger diameter.

14. Select the Front plane, then pick the Sketch tool.

15. Change the part to Hidden Lines Removed in the View Heads Up Toolbar.

16. Draw a horizontal Centerline through the origin.

17. Use the Corner Rectangle tool to draw a small rectangle coincident with the top edge of the 1.25-inch diameter shaft as shown below.

18. Use the Smart Dimension tool to size and locate the small rectangle. (Note that the 1.17-inch value is a diameter dimension.)

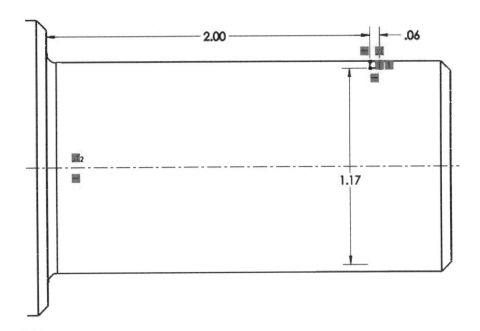

19. Exit Sketch.

20. Pick the Revolved Cut icon to revolve the section about the centerline through 360 degrees. Pick the green checkmark. (Note that the Revolved Cut icon is hidden behind the Extruded Cut icon.)

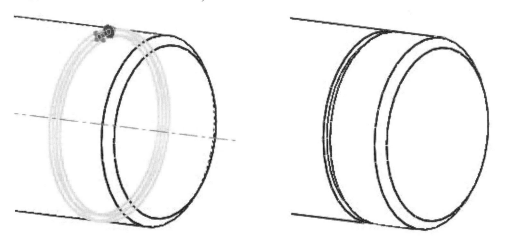

The finished step shaft still needs its material selected. In the description, it stated that it was made from AISI 1045 cold-drawn steel. When we look in the list of common materials, this material is not listed so we need to search deeper into the material table.

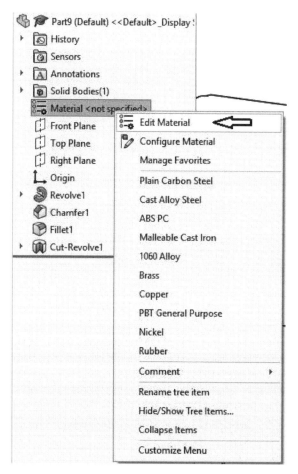

21. Move the cursor to **Material <not specified>** in the Feature Manager Design tree. Right-click to get a pop-up menu to appear. Pick **Edit Material** with the LMB to bring up the Material window. Under Steel, select **AISI 1045 Steel, cold drawn** from the list to assign this material to this part. In the Units area, change the units to English (IPS). Pick the **Apply** button. Pick **Close**.

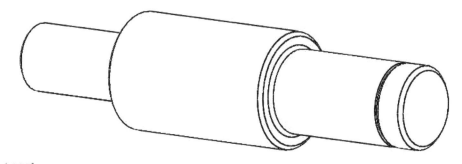

22. Change the part to **Shaded** in the View Heads Up Toolbar.

23. Save the Part by picking this icon or under **File> Save**.

24. Make sure you are saving the part in the SolidWorks Part folder. Name the part file **"Step Shaft"**, then pick the **Save** button.
25. Select **File> Print...**, (if asked to).
26. Select **File> Close** or type <Ctrl>+W to close the part and clear the window.

Revolve Boss/Base Exercise

A clutch is needed for a 15-horsepower one-cylinder, two-cycle engine that develops its maximum torque at 3400 r.p.m. Design a cone clutch with a maximum outside diameter of 7 inches and a cone angle of 12°. (Typical cone angles are 10°, 12°, and 15°.) Assume the coefficient of friction between the cone and the cup material is 0.32 and the maximum allowable pressure that can be exerted on the braking material is 50 psi. Assume a 1.00-inch diameter shaft can handle the maximum engine torque. Assume a minimum safety factor of 1.5.

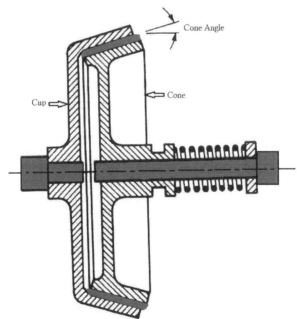

Assumptions:

   Shaft diameter = 1.00 inches
   Cone/cup angle = 12 degrees
   Brake material average coefficient of friction, $f = 0.32$

Uniform wear on brake material
Brake material thickness on cone = 0.125 inches
Cone Large diameter, Dia$_{cone}$ = 6.00 inches
Width of brake material in contact with the clutch cup = 1.00 inches
The minimum factor of safety for the clutch = 1.5

Calculations lead to:

Cone's material large diameter in contact with the cup,

$$D = Dia_{cone} + 2 \cdot (Thickness_{Material}) \cos(Angle_{Cone}) = 6.245 \; inches$$

Cone's small diameter at contact,

$$dia_{Cone} = Dia_{cone} - 2 \cdot Width_{Cone} \cdot \sin(Angle_{Cone}) = 5.584 \; inches$$

Cone's material small diameter at contact with the cup

$$d = D - 2 \cdot Width_{Cone} \cdot \sin(Angle_{Cone}) = 5.829 \; inches$$

From H = 30 hp at n = 3400 r.p.m, thus the maximum Torque (T) is,

$$H = \frac{T \cdot n}{63025} \rightarrow \; T = 556 \; in.lbs.$$

Maximum pressure (P_a) on braking material must be less than 50 psi,

$$P_a = \frac{8 \cdot T \cdot \sin(Angle_{Cone})}{\pi \cdot f \cdot d(D^2 - d^2)} = 31.5 \; psi < 50 \; psi \; (S.F. = \frac{50}{31.5} = 1.59)$$

The maximum Force (F) to be applied by the spring is,

$$F = \frac{\pi \cdot P_a \cdot d(D^2 - d^2)}{2} = 120 \; lbs.$$

Design OK.

More assumptions:
Diameter of the cone's hub = 2.00 inches
Length of the cone's hub = 2.24 inches
The cone's hub is splined so it can slide on the shaft and transmit torque
Width of shifting groove = 0.250 inches
Depth of shifting groove = 0.200 inches (I.D. = 1.60 inches)
Diameter of the cup's hub = 2.00 inches
Length of the cup's hub = 1.25 inches
The cup's hub is keyed to the shaft
The cup is locked in place on the shaft with a setscrew opposite the keyway

Now that the design is complete, let's create the cone for the clutch.

1. Let's begin by starting SOLIDWORKS. Be sure to toggle into **Advanced** mode. Select the **Part_IPS_ANSI** SOLIDWORKS Template, then pick **OK**.

2. Pick the Front plane from the **Feature Manager Design Tree**, then select the sketch tool.

3. Draw a horizontal centerline through the origin.

4. Use the Line tool to create the shaft profile above the horizontal centerline as shown on the following page.

5. Pick the centerline then hold down the <Ctrl> key and pick the origin. In the Add Relations window, pick Coincident. Make the left edge of the clutch cup directly above the origin.

6. Use the Smart Dimension tool to dimension the sketch as shown on the following two pages.

Adding a diameter dimension is done by picking the centerline and the diameter point with the LMB, and then moving the mouse cursor below the centerline and pressing the LMB to place the diameter dimension at the current cursor position. You can also pick the diameter point first, then the centerline followed by moving the cursor below the centerline to place the dimension.

Note that the faded 1.142-inch dimension is a driven dimension used as a reference.

7. Exit Sketch.

8. With **Sketch 1** highlighted, pick the Revolved Boss/Base icon.

9. With Blind and 360 degrees set, pick the green checkmark.

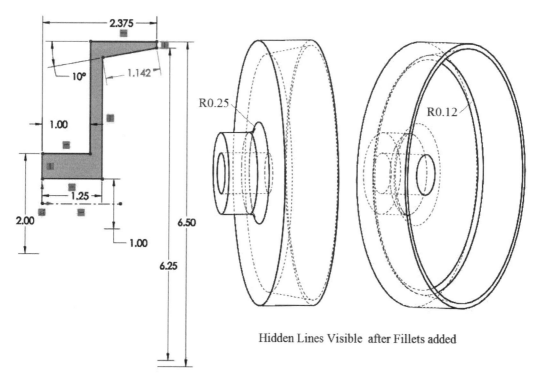

Hidden Lines Visible after Fillets added

To lower the stresses at two critical locations, add a 0.25-inch fillet where the hub meets the larger diameter and another 0.12 fillet where the inner hub meets the vertical wall.

10. Use the Fillet tool to add a 0.25-inch round where the hub meets the larger diameter.

11. Use the Fillet tool to add a 0.12-inch round where the inner hub meets the vertical wall.

Add a keyway to allow for the transmission of torque using a square key in the 1.00-inch diameter hole. From the Table, a 1.00-inch diameter shaft should use a ¼ inch square key.

12. Pick the Sketch icon, then pick the flat face of the hub.

13. Draw a vertical centerline through the origin.

14. Use the Center Rectangle to draw a rectangle at the top of the 1.00-inch diameter hole.

15. Add a Point to the bottom of the 1.00-inch diameter hole. This is needed so a dimension can be added between the top of the rectangle and the bottom of the hole.

16. Use Smart Dimension to add the two appropriate dimensions.

Table 4-1 Recommended Key Sizes

Shaft Diameter		Key Size		Keyway
Over (inch)	Including (inch)	w (inch)	h (inch)	Depth (inch)
5/16	7/16	3/32	3/32	3/64
7/16	9/16	1/8	3/32	3/64
		1/8	1/8	1/16
9/16	7/8	3/16	1/8	1/16
		3/16	3/16	3/32
7/8	1 1/4	1/4	3/16	3/32
		1/4	1/4	1/8
1 1/4	1 3/8	5/16	1/4	1/8
		5/16	5/16	5/32
1 3/8	1 3/4	3/8	1/4	1/8
		3/8	3/8	3/16
1 3/4	2 1/4	1/2	3/8	3/16
		1/2	1/2	1/4
2 1/4	2 3/4	5/8	7/16	7/32
		5/8	5/8	5/16
2 3/4	3 1/4	3/4	1/2	1/4
		3/4	3/4	3/8

Keyway Calculations

D = diameter of the hole = 1.00 inch

W = width of the key or keyway = 0.25 inch

H = distance from the top of the keyway to the opposite side of the hole

$$H = \frac{D + W + \sqrt{D^2 - W^2}}{2} = \frac{1.00 + 0.25 + \sqrt{1.00^2 - 0.25^2}}{2} = 1.109 \; inches$$

Let's use 1.115 inches for this distance which gives us a small clearance at the top of the keyway.

17. [icon] Exit Sketch.

18. [icon] Use the **Extruded Cut** to cut the keyway **Through All**.

19. Pick the green checkmark to accept the extruded cut.

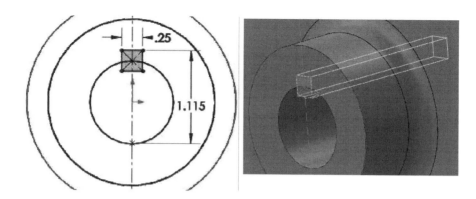

20. Use the Fillet tool to add 0.03-inch rounds to the remaining sharp edges.

21. Use the Chamfer tool to add 0.06-inch x 45° chamfer to both ends of the 1.00-inch diameter hole.

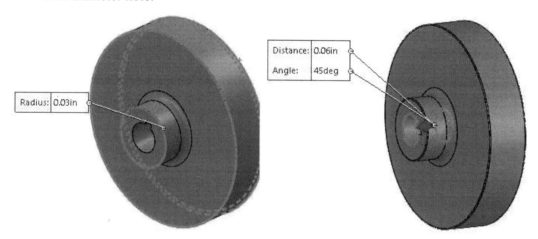

22. Save the part in the SolidWorks Parts folder. Name it "**Clutch Cup 15 hp**".

23. Use the **Hole Tool** to create a threaded hole 0.50 inches from the front face of the hub and aligned with the Front plane. This should place the threaded hole directly opposite the keyway.

24. For Hole Type, select the **Straight Tap** icon. Set Standard to **ANSI** and Type to **Tapped Hole**. See the figure on the following page.

25. Under Hole Specifications, select ¼-20 from the list.

26. For End Condition and Thread depth, select **Up To Next**.

27. Under Precision, select **.123 (Document).**

28. Pick the Positions tab.

29. Use the LMB to pick the outer surface of the hub as the location of the hole.

30. Use the LMB to pick a point on the outer surface approximately 0.50 inches from the front face of the hub.

31. Select the Smart Dimension tool, then pick the center point of the hole. In the Control Vertex Parameters, enter X = 0.5, Y =-1.0, and Z = 0.0.

32. Pick the green checkmark to complete the process.

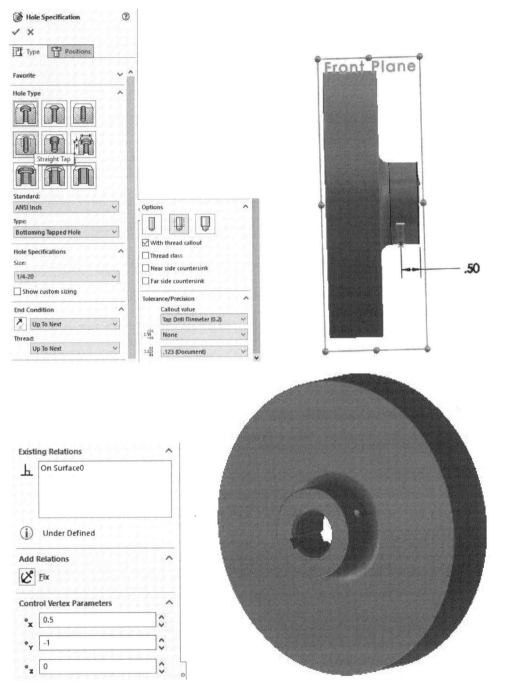

33. Right-click on **Material \<not specified\>** in the **Feature Manager Design Tree**, then pick **Plain Carbon Steel** from the list.

34. 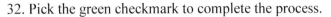 Save the Part again by picking the Save icon or under **File\> Save**.

35. Select **File\> Print...** (if asked to)

36. Select **File\> Close** or type \<Ctrl\>+W to close the part and clear the window.

Revolved Boss/Base Problems

4-1   Design the **"Clutch Cone 15hp"** for the 15 hp clutch made from plain carbon steel as described below. You can leave the spline out of the hole at this time. We will add the spline as one of the problems in the next chapter which covers patterns. Add a 0.06-inch x 45° chamfer to both ends of the center hole. Break all sharp edges using a 0.03-inch fillet. Note that the (5.88) diameter is a driven reference dimension. **Design change:** change the 0.25-inch wide slot to 0.26 inches.

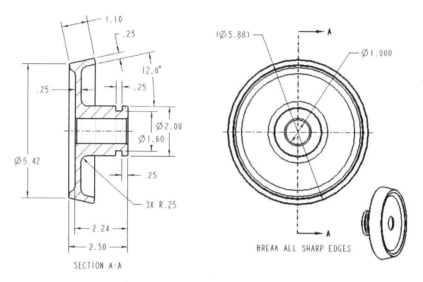

4-2   Design the **"Step Pulley"** shown below made from AISI 1020 cold-rolled steel. The origin should be on the centerline at the left end of the pulley. The 3.00-inch diameter hole has a 0.06-inch x 45° chamfer at both ends. **Design change:** increase the 7.00-inch inside diameter to 7.02 inches.

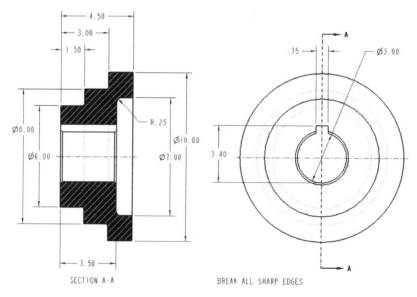

4-3   Design the **"Nozzle"** made from ABS PC. The origin needs to be on the centerline at the left end of the nozzle. **Design change**: decrease the 0.63-inch diameter to 0.060 inches. What is the weight of the nozzle?

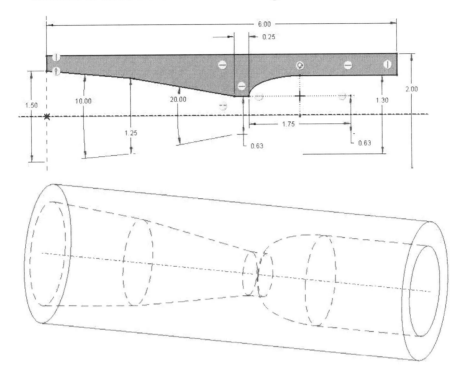

4-4   Design the **"Axle"** made from plain carbon steel. The origin needs to be on the centerline at the location where the diameter changes from 1.1178 inches to 0.7871 inches. The left groove has a diameter of 0.875 inches. The preferred method is to create the Axle first, then use revolved cuts to create the grooves, then add the chamfers and fillet last. When designing this axle we will use its MMC (Maximum Material Condition).

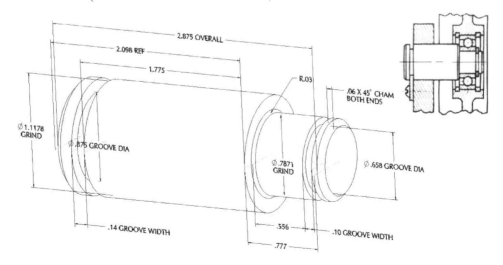

4-5 Design the **"Offset Shaft"** made from AISI 4340 annealed steel. The origin is on the centerline at the left end of the part. The outside diameter of the 1.00-inch diameter shaft is tangent with the outside diameter of the 1.50-inch diameter portion. **Design change:** increase the depth of the 1.25-inch cutout from 0.25 inches to 0.28 inches. What is the weight of this offset shaft?

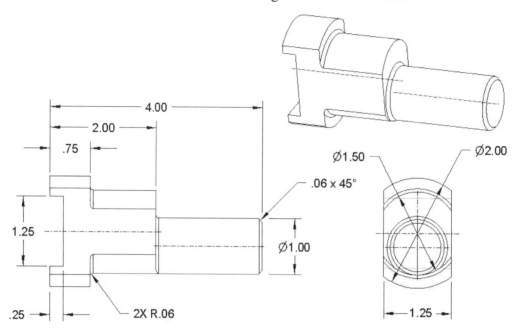

4-6 Design the **"V-belt Pulley-6.0"** made from malleable cast iron. The origin is on the centerline at the front of the hub. Add a 0.03-inch x 45° chamfer to both ends of the center hole. **Design change:** increase the width of the keyway to 0.252 inches.

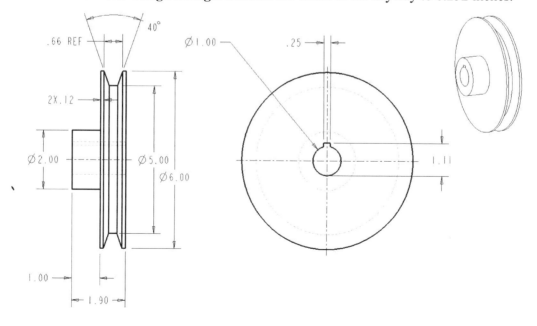

4-7   Design the **"V-belt Idler"** made from AISI 1010 hot rolled bar steel. The counter bore is applied to both ends of the idler pulley to allow for roller bearings to be inserted. **Design change:** increase the depth of the counterbores from 0.50 inches to 0.52 inches.

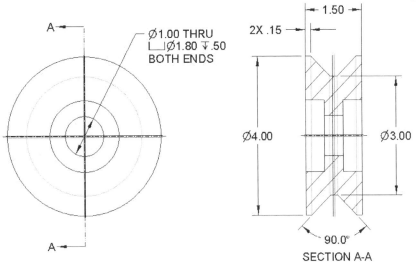

Ø1.00 THRU
⌴Ø1.80 ⬇.50
BOTH ENDS

2X .15

1.50

Ø4.00

Ø3.00

90.0°

SECTION A-A

4-8   Design the **"Rod Head"** made from 1060 alloy. The origin is on the centerline at the left end of the part. **Design change:** decrease the 13-mm diameter hole to 10 mm. All Dimensions are in millimeters. What is the mass of this part?

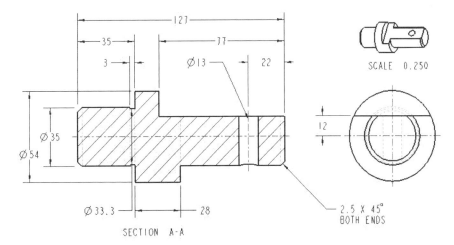

127

35

3

Ø13

77

22

SCALE   0.250

12

Ø 35

Ø 54

Ø 33.3

28

2.5 X 45°
BOTH ENDS

SECTION   A-A

4-9   Design the **"Tapered Roller"** made from malleable cast iron. The three tapered sections are at 15 degrees from the horizontal. How far does the 0.75-inch diameter reach into the Tapered Roller from the outer left edge of the .75-inch hole? _____ inches (0.xxx)

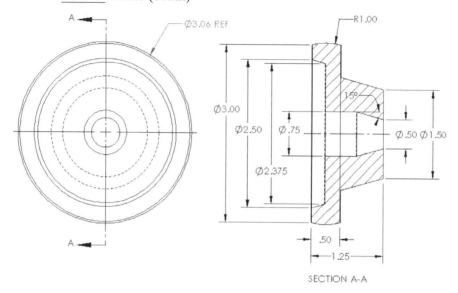

SECTION A-A

4-10 Design a 0.75-inch diameter **"Tapered Shaft"** that is 6 inches long with the same 15-degree tapers so that the Tapered Roller shown above can be pressed onto this shaft. The shaft is made from plain carbon steel. Add a ¼-20 UNC threaded hole 0.50 inches deep on the tapered end. (See problem 4.9) What is the horizontal distance for the required 15° taper? _____ inches (0.xxx)

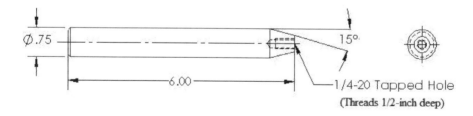

4-11 Design the **"Shaft Bracket"** made from plain carbon steel. All fillets and rounds are .06-inches.

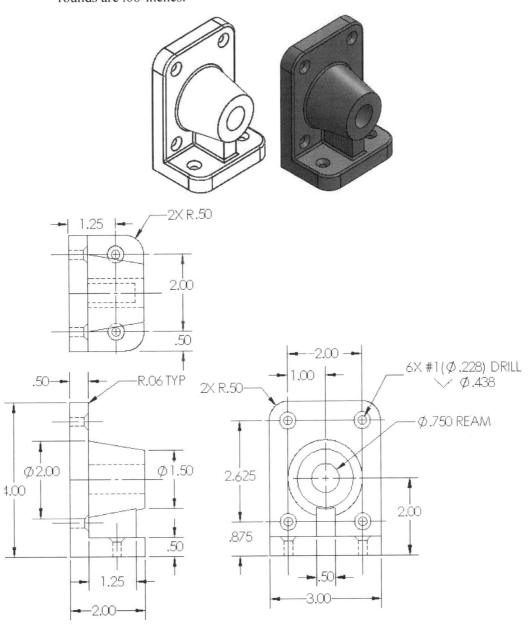

4-12 Design the **"Socket"** made from AISI 1020 steel. Note that the center hole is tapered as well. It has a .75-inch diameter at the top and a .50-inch diameter at the bottom. Dimensions enclosed in parenthesis are reference dimensions.

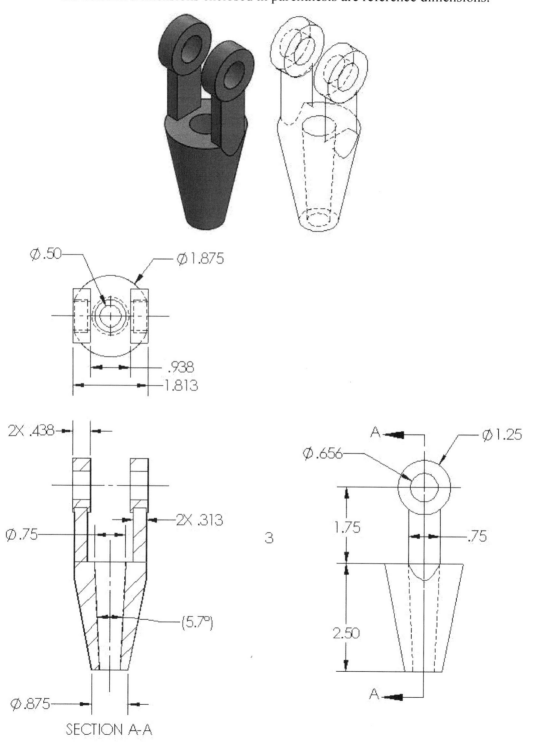

SECTION A-A

Chapter 5.   Patterned and Mirrored Features

Introduction

Parts often contain repeated features of the same size and shape. In SOLIDWORKS these repeated features are called patterns. The first instance of the pattern is called the pattern leader. Not only can patterns duplicate the exact size and shape of a feature, but they can change the size and shape at each instance along the way. Feature patterns can be one-dimensional or two-dimensional. There are seven types of feature patterns available. They are Linear, Circular, Curve Driven, Sketch Driven, Table Driven, Fill, and Variable. Besides these patterns, there is also a Mirror command which will be discussed next.

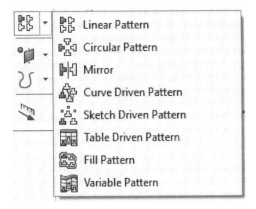

The **Mirror** command allows the part or a feature of the part to be mirrored across a given axis or plane. In this case, Boss-Extrude1 is mirrored about the RIGHT plane.

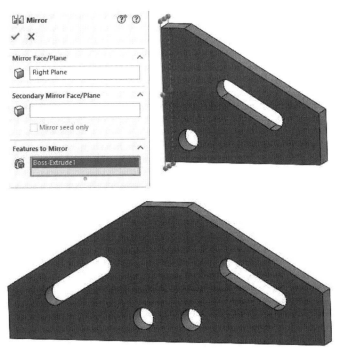

An individual feature can be mirrored as well. In this case, just the slot is mirrored about the RIGHT Plane. Also, note that if the original feature changes size, then the mirrored feature will reflect that change.

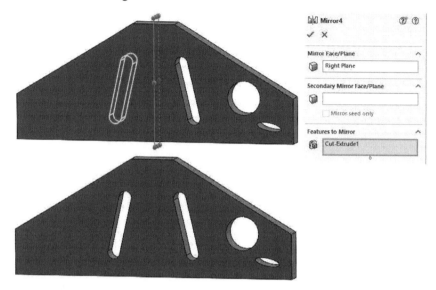

The feature can also be mirrored about two different planes as shown below.

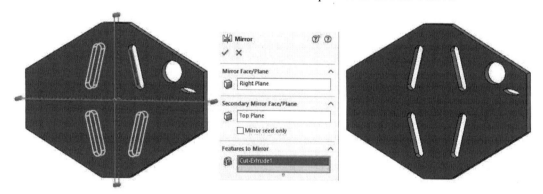

Let's look at the **Linear** pattern. After picking the linear pattern icon, select the edge, face, or axis to define Direction 1 of the pattern. Next, select the feature to be patterned from the model geometry. Enter the spacing and the number of incidents, (0.95 inches and 4). Pick the green checkmark to accept the pattern.

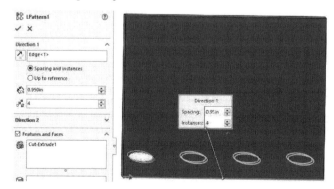

If the pattern is to be two-dimensional, then select the second edge, face, or axis for Direction 2 along with the spacing and number of incidents for the second direction, (1.00 inches and 3). Pick the green checkmark to accept the pattern.

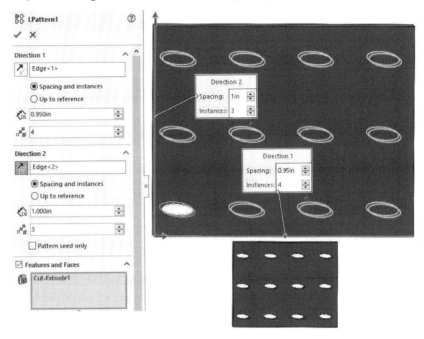

If you want to suppress an occurrence of the pattern, expand the Instance to Skip section in the Linear Pattern window, then select using the LMB the patterns you wish to disable. In this case, the two inner patterns were suppressed.

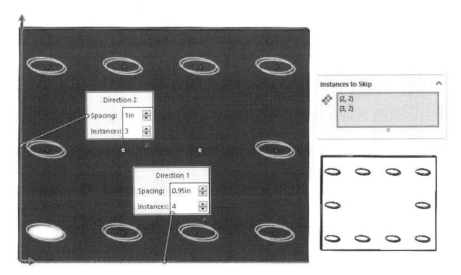

The **Circular** pattern creates patterns around an axis. After picking the Circular pattern icon, click on the Pattern Axle selection box in Direction 1, then select the axis or cylindrical surface that the pattern is to be rotated about. If you want equal spacing, select that option along with 360 degrees and the number of incidents. In the Features and Faces area, pick the feature you want to rotate about the given axis or cylindrical surface. Pick the green checkmark to complete the process.

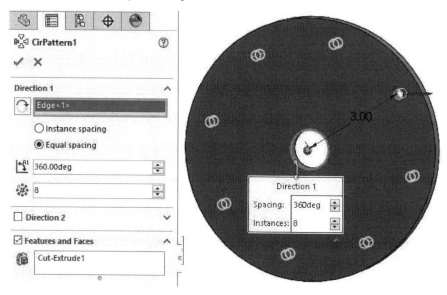

If you want to set the angle between patterns, then select the Instance spacing option followed by the angle-spacing between patterns and the number of incidents. In this case, 10 incidents spaced 20 degrees apart.

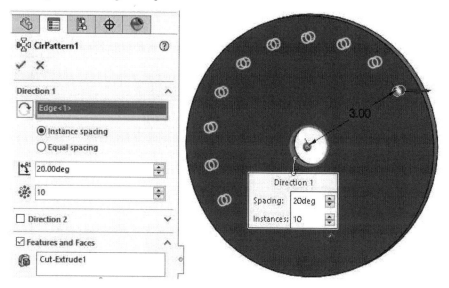

If you want to suppress an occurrence of the pattern, expand the Instance to Skip section in the Circular Pattern window, then using the LMB select the patterns you wish to disable. In this case, the third and eighth patterns were suppressed. This feature works for all the different pattern types.

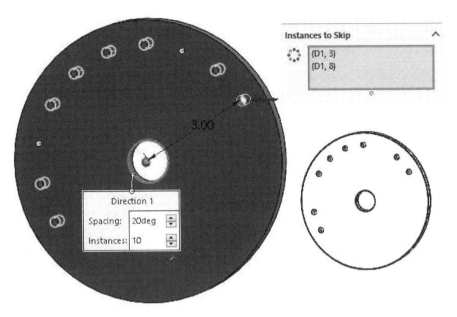

When placing holes in a racetrack-style gasket, the **Curve Driven** pattern works best. Activate the Curve Driven pattern, then select the curve it is to follow (**Sketch3** in this case). Enter the number of incidents and check the Equal spacing box. Select the feature to pattern (**Cut-Extrude1** in this case), then pick the green checkmark to complete the process. Note that Boss-Extrude1 is the gasket, Cut-Extrude1 is the hole, and Sketch3 is the curve to follow.

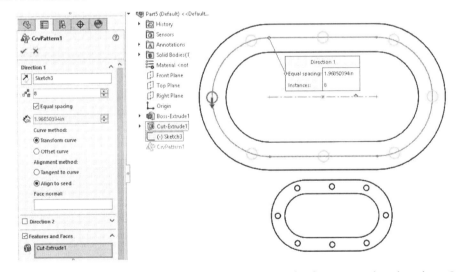

The **Sketch Driven** pattern allows for the placement of a feature at sketch points. It is the same procedure as the Curve Driven pattern, except the feature is copied to each sketch point in the assigned sketch.

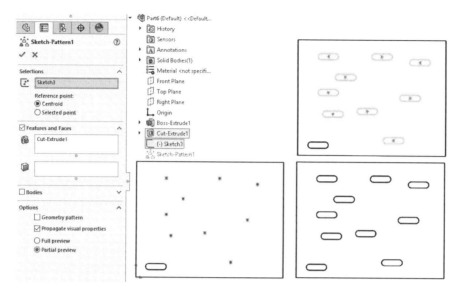

The **Fill Pattern** is used to populate an area or a surface with repeated features. This feature requires the selection of the surface or sketched area along with the spacing between features, the angle the features move in, and the general direction of the pattern. In this case, Cut-Extrude1 is patterned over the entire surface with 0.50 inches between patterns with the patterns moving at a 60-degree angle.

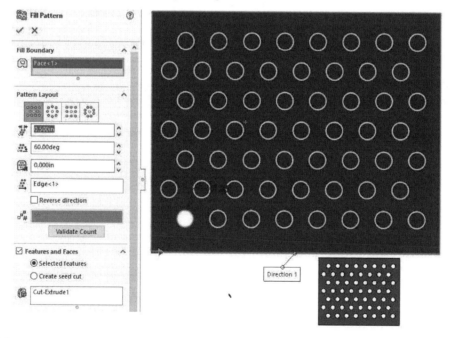

The fill pattern can also be circular, square, or polygon-shaped as indicated by the icon in the Fill pattern window.

Patterned Features Practice (Linear)

The picture below is an antique erector set plate with many holes. If you were to create this part in SOLIDWORKS and individually create each hole in the plate, it would take a long time. We can use the linear pattern feature of SOLIDWORKS and quickly create most of the holes.

Design Intent: Create the erector set plate to scale. The plate is 4.00 inches long and 3.00 inches wide, made from 0.035-inch thick plain carbon steel. The holes are 0.166 inches in diameter (#19 DRILL) and spaced 1/2 inch apart in both directions. The single hole on the left end is 3.75 inches away from the right end and centered on the plate. The two edges are bent over creating a 1/2-inch lip with the same hole spacing. Assume the origin is in the middle of the plate, thus a Mid Plane extrusion is required.

1. Let's begin by starting SOLIDWORKS. Be sure to toggle into **Advanced** mode. Select the **Part_IPS_ANSI** SOLIDWORKS Template, then pick **OK**.

2. If the Sketch and Features tools are not on the sides of the window, you can also add the Sketch and Features tools to the sides of the window. The Sketch tools will appear along the right side of the window. The Features tool will appear along the left side of the window. These are added by **Toolbars** from the same pop-up menu.

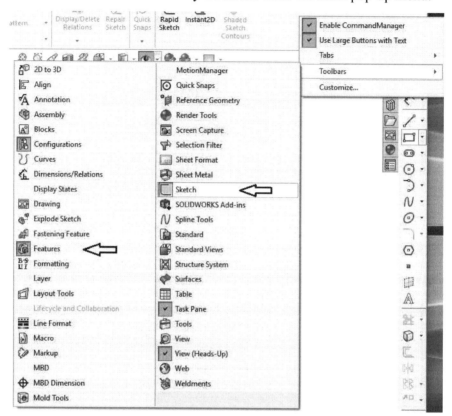

3. Selecting the **CommandManager** from the pop-up menu disables it. Picking it again enables it. Disabling the **Use Large Buttons with Text** option gives you more room to work.

We will create the erector plate by sketching its 2D cross-section, and then extruding it to a length of 4 inches.

4. Pick the RIGHT plane from the **Feature Manager Design Tree**, then select the sketch tool.

5. Draw a vertical centerline through the origin.

6. Use the Line tool to sketch the cross-section of the erector plate symmetric about the origin as shown below.

7. Use the Smart Dimension tool to dimension the sketch.

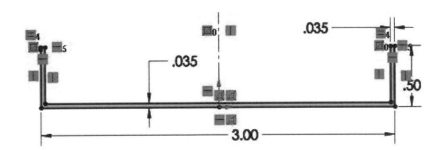

8. 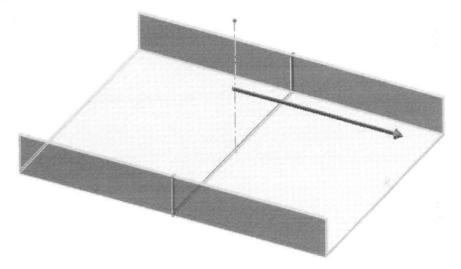 Exit Sketch.

9. With **Sketch1** highlighted, select the Extruded Boss/Base icon. For Direction 1, select Mid Plane from the pull-down menu. Set the length to 4.00 inches. Pick the green checkmark.

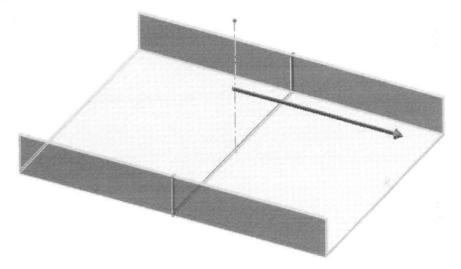

10. Select the flat portion of the plate, then pick the Sketch icon.

11. Use the Circle tool to create a small circle in the lower right corner of the plate as shown.

12. Use the Smart Dimension tool to size and locate the small hole.

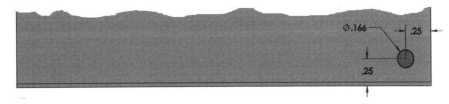

13. Exit Sketch

14. With **Sketch2** highlighted, select the Extruded Cut icon. Select **Through All** for Direction 1. Pick the green checkmark to create the first hole.

15. With **Cut-Extrude1** highlighted, select Linear Pattern from the Extruded Boss/Base Toolbar.

16. For Direction 1, select the lower horizontal edge of the plate. Set the distance between patterns to 0.500 inches and the number of incidents to 7.

17. For Direction 2, select the right vertical edge of the plate. Set the distance between patterns to 0.500 inches and the number of incidents to 6.

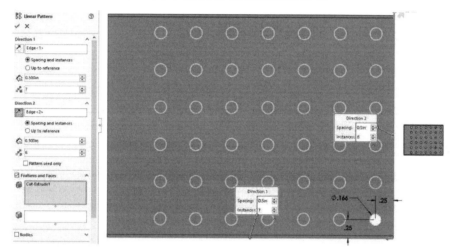

18. Pick the green checkmark to complete the linear pattern process.

19. Select the flat portion of the plate, then pick the Sketch icon.

20. Use the Circle tool to create a small circle near the far left edge of the plate as on the following page.

21. Use the Smart Dimension tool to size and locate the small hole.

22. Exit Sketch

23. With **Sketch3** highlighted, select the Extruded Cut icon. Select **Through All** for Direction 1. Pick the green checkmark to create this single hole.

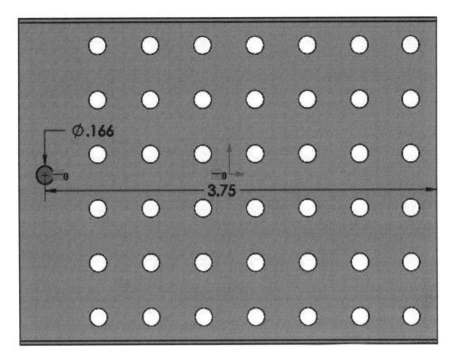

24. From the View Orientation icon, pick the Front view.

25. 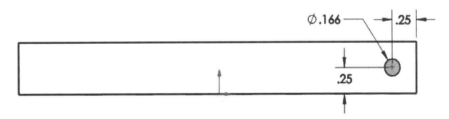 Select the side surface, then pick the sketch icon.

26. Use the Circle tool to create a small circle near the right end of the surface as shown below.

27. Use the Smart Dimension tool to size and locate the small hole.

28. Exit sketch

29. With **Sketch4** highlighted, select the Extruded Cut icon. Select **Through All** for Direction 1. Pick the green checkmark to create this single hole through both sides.

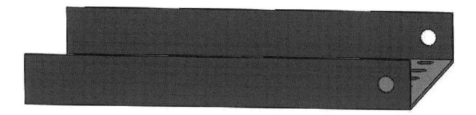

30. 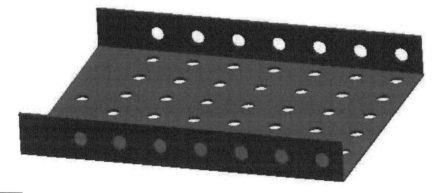 With **Cut-Extrude3** highlighted, select Linear Pattern from the Extruded Boss/Base Toolbar.

31. For Direction 1, select the lower horizontal edge of the plate. Set the distance between patterns to 0.500 inches and the number of incidents to 7.

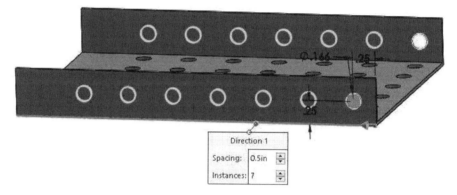

32. Pick the green checkmark to complete the linear pattern process.

33. Right-click on **Material <not specified>** in the **Feature Manager Design Tree**, then pick **Plain Carbon Steel** from the list.

34. Select the **Edit Appearance** icon from the View Heads Up Toolbar. In the color chart, pick red, then pick the part in the graphics window. Pick the green checkmark.

35. Save the Part by picking this icon or under **File> Save**.

36. Name the part file "**Erector Set Plate 3x4**", then pick the **Save** button.

37. Select **File> Print...** (if asked to)

38. Select **File> Close** or type <Ctrl>+W to close the part and clear the window.

Patterned Features Practice (Circular)

In this practice session, we will design a hollow-ground, 7.25-inch diameter circular saw blade. Cutting the teeth on the saw blade will be done using the Circular Pattern tool.

Design Intent: Design a 7.25-inch diameter hollow ground, 72-tooth saw blade with a 0.625-inch diameter hole for mounting. The hollow ground saw blade is used for satin-smooth crosscuts, rips, and miters in wood. The cutting portion of the blade is 0.056 inches thick. Its hub is 2.50 inches in diameter and 0.100 inches thick. The saw blade is made from AISI 347 Annealed Stainless Steel (SS). In the figure below, a small 1/4-inch hole was added to the saw blade to balance it for high-speed operation. We will not create this hole.

1. Let's begin by starting SOLIDWORKS. Be sure to toggle into **Advanced** mode. Select the **Part_IPS_ANSI** SOLIDWORKS Template, then pick **OK**.

We will create the saw blade blank by sketching its 2D cross-section, then revolving it around its horizontal centerline.

2. Pick the RIGHT plane from the Feature Manager Design Tree, then select the sketch tool.

3. Draw a horizontal centerline through the origin.

4. Use the Line tool to sketch the cross-section of the saw blade blank symmetric about the origin as shown below. Note that the two small horizontal lines at the 2.50-inch diameter are set equal in length to ensure the saw blade is symmetrical.

5. Use the Smart Dimension tool to dimension the sketch.

6. Exit Sketch

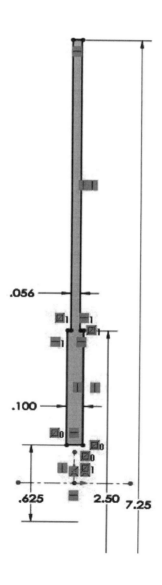

.056

.100

.625   2.50 7.25

7. With **Sketch1** highlighted, select the Revolve Boss/Base icon.

8. Pick the green checkmark to complete the revolved section through 360 degrees.

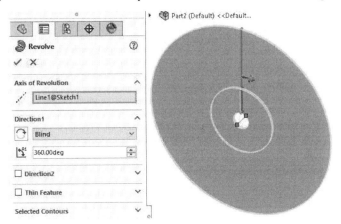

9. Select the large diameter of the saw blade blank, then pick the Sketch icon.

10. Pick the Hidden Lines Removed from the View Heads Up Toolbar.

11. Using the centerline tool, sketch a horizontal centerline along with two centerlines at approximately 20 degrees and 25 degrees as shown below. Since there are 72 teeth, each tooth uses 5 degrees, thus the 5-degree difference in the centerlines.

12. Use the Smart Dimension tool to make sure the two centerlines are positioned correctly.

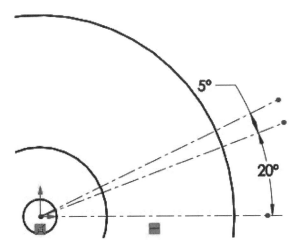

13. Use the Line tool to draw two lines that will represent the tooth cutout.

14. Using the Convert Entities from the right side of the window, pick the outer diameter circle of the saw blank.

15. Use the Trim Entities tool to remove the portion of the large circle that is not within the tooth cutout. The tooth cutout should now be shaded showing that it is a closed area.

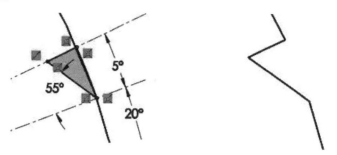

16. Exit Sketch

17. Use the Extruded Cut tool. For Direction 1, select **Through All**. Pick the green checkmark. See the figure above.

18. Highlight **Cut-Extrude1**, then pick the Circular Pattern icon. Pick the center hole of the saw blade as Direction 1. Select Equal spacing. Enter 72 for the number of incidents. Pick the green checkmark.

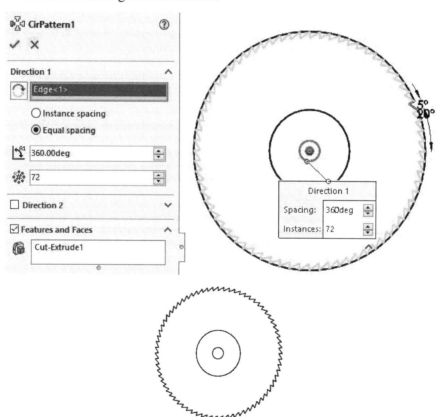

19. 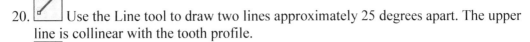 Select the saw blade surface, then pick the Sketch icon.

20. Use the Line tool to draw two lines approximately 25 degrees apart. The upper line is collinear with the tooth profile.

21. Use the Smart Dimension tool to force the two lines to be 25 degrees apart.

22. Use the 3-Point Arc between the two lines. Add sketch relations that force the arc to be tangent with the two lines.

23. Use the Smart Dimension tool to give the arc a .12-inch radius.

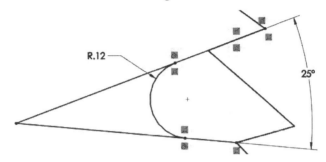

24. Use the Trim Entities tool to remove the left side unwanted intersecting lines.

25. Use the 3-Point Arc tool and create an arc that goes through the 3 points shown. The area should become shaded since it is now a closed area.

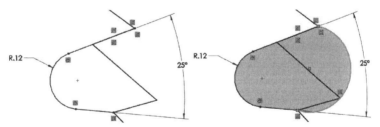

26. Exit Sketch

27. Use the Extruded Cut tool to remove the material in the shaded area. Set Direction 1 to **Through All**. Pick the green checkmark.

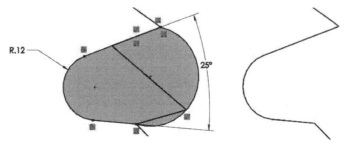

28. 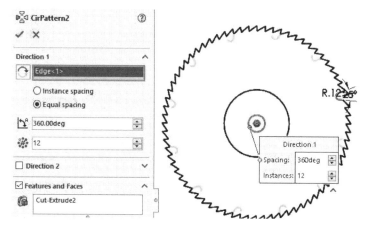 With **Cut-Extrude2** highlighted, select the Circular Pattern tool.

29. Pick the center hole as Direction 1. Set the number of incidents to 12, equally spaced over 360 degrees.

30. Pick the green checkmark to complete the circular pattern process.

This completes the circular saw blade construction.

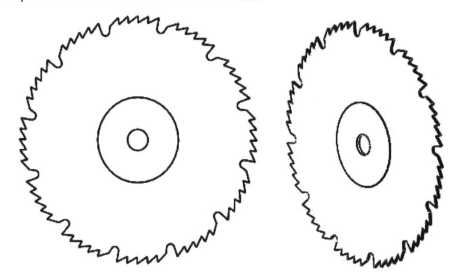

31. Right-click on **Material <not specified>** in the Feature Manager Design Tree, then pick Edit Material. In the Material window, select **AISI 347 Annealed Stainless Steel (SS)** from the list. Pick **Apply**. Pick **Close**.

32. Save the Part by picking the Save icon or at the top select **File> Save**.

33. Name the part file "**Circular Saw Blade**", then pick the **Save** button.

34. Select **File> Print…** (if asked to)

35. Select **File> Close** or type <Ctrl>+W to close the part and clear the window.

Patterned Features Exercise (Circular)

The dual-toothed sprocket shown below is to be designed. We will use the Revolved Boss/Base tool to create the sprocket blank and the Circular Pattern tool to duplicate the sprocket teeth. We will use the Extruded Cut tool to cut the keyway.

Design Intent: Design the dual-toothed, 17-tooth #40 and 21-tooth #50 sprocket. The smaller sprocket is 3.00 inches in diameter and 1/4 inches thick. The larger sprocket is 4.50 inches in diameter and 5/16 inches thick. The bore is 1.25 inches in diameter with the outside hub diameter equal to 1.75 inches. The hub protrudes past the larger sprocket by 1/4 inch so that the larger sprocket can be welded to the hub. The smaller sprocket is welded to the hub on the inner side. The keyway is 5/16 inches wide with a distance of 1.36 inches from the top of the keyway to the opposite side of the hole. Because the hub is only 1.75 inches in diameter, the keyway is not going to be cut to its standard depth. Place the origin at the outer surface of the smaller sprocket and the center of the hole.

We need to do some calculations before we begin creating this 3D model.

For the #40 sprocket with 17 teeth and a roller diameter of 0.312 inches:

$$pitch = p = 40/80 = 0.500 \ inches$$

$$Outside\ Diameter = OD = 0.500 * \left(0.6 + \frac{1}{\tan(\frac{180}{17})} \right) = 2.975\ inches$$

$$PitchDiameter = D_{17} = \frac{0.500}{\sin\left(\frac{180^o}{17}\right)} = 2.721\ inches$$

For the #50 sprocket with 21 teeth and a roller diameter of 0.400 inches:

$$pitch = p = 50/80 = 0.625 \; inches$$

$$Outside \; Diameter = OD = 0.625 * \left(0.6 + \frac{1}{\tan(\frac{180}{21})} \right) = 4.522 \; inches$$

$$PitchDiameter = D_{21} = \frac{0.625}{\sin\left(\frac{180^o}{21}\right)} = 4.193 \; inches$$

1. Let's begin by starting SOLIDWORKS. Be sure to toggle into **Advanced** mode. Select the **Part_IPS_ANSI** SOLIDWORKS Template, then pick **OK**.

We will create the dual-tooth sprocket blank by sketching its 2D cross-section, and then revolving it around its horizontal centerline.

2. Pick the RIGHT plane from the **Feature Manager Design Tree**, then select the sketch tool.

3. Draw a horizontal centerline through the origin.

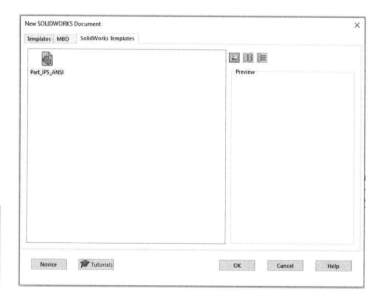

4. Use the Line tool to sketch the cross-section of the dual-toothed sprocket.

5. Use the Smart Dimension tool to dimension the cross-section according to the following figure.

6. Exit Sketch.

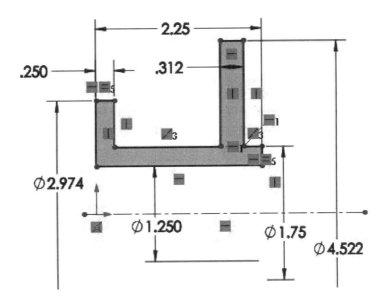

7. 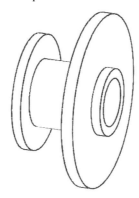 With **Sketch1** highlighted, select the Revolve Boss/Base icon.

8. Pick the green checkmark to complete the revolved section through 360 degrees.

9. Pick the surface of the smaller sprocket blank, then pick the Sketch icon.

10. Draw a horizontal centerline and a second centerline at a 30° angle.

11. Draw a circle centered at the origin with a diameter of 2.721 inches.

12. Right-click on the circle, then in the pop-up menu pick the Construction Geometry icon to convert the circle to a construction feature.

13. Draw a circle at the intersection of the 30° centerline and the construction circle. Set its diameter to 0.312 inches (0.156-inch radius).

14. Use the Line tool to draw a horizontal line tangent to this small circle, and a second line tangent to the small circle at a 60° angle.

15. Select the Convert Entities tool, then pick the outer edge of the smaller sprocket blank. Pick the green checkmark to copy this circle to the current sketch.

16. Use the Trim Entities to remove the portion of the circle outside the cut area. Also, trim the portions of the small circle inside the shaded area. The cut area should now be shaded as shown below.

17. Exit Sketch.

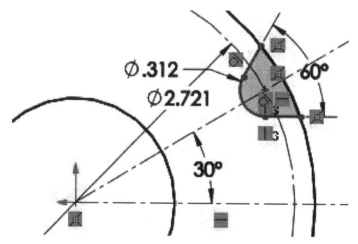

18. With **Sketch2** highlighted, pick the Extruded Cut icon. In the Direction 1 window, pick **Up to Next** in the pulldown menu. Pick the green checkmark.

19. With Cut-Extrude1 highlighted, pick the Circular Pattern icon. Pick the center hole of the sprocket as Direction 1. Set the number of incidents to 17 equally spaced over 360 degrees. Pick the green checkmark.

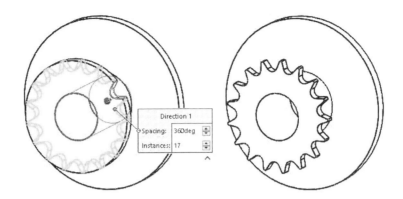

20. Pick the surface of the large sprocket blank, then pick the Sketch icon.

21. Draw a horizontal centerline and a second centerline at a 30° angle.

22. Draw a circle centered at the origin with a diameter of 4.193 inches.

23. Right-click on the circle, then in the pop-up menu pick the Construction Geometry icon to convert the circle to a construction feature.

24. Draw a circle at the intersection of the 30° centerline and the construction circle. Set its diameter to 0.400 inches (0.200-inch radius).

25. Use the Line tool to draw a horizontal line tangent to this small circle, and a second line tangent to the small circle at a 60° angle.

26. Select the Convert Entities tool, then pick the outer edge of the larger sprocket blank. Pick the green checkmark to copy this circle to the current sketch.

27. Use the Trim Entities to remove the portion of the circle outside the cut area. Also, trim the portions of the small circle inside the shaded area. The cut area should now be shaded as shown below.

28. Exit Sketch.

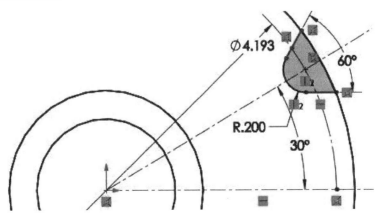

29. With **Sketch3** highlighted, pick the Extruded Cut icon. In the Direction 1 window, pick **Up to Next** in the pulldown menu. Pick the green checkmark.

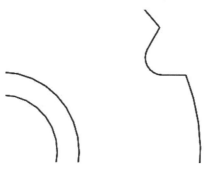

30. 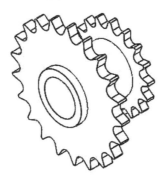 With Cut-Extrude2 highlighted, pick the Circular Pattern icon. Pick the center hole as Direction 1. Set the number of incidents to 21 equally spaced over 360 degrees. Pick the green checkmark.

31. Pick the outer surface of the smaller sprocket, then pick the Sketch icon.

32. Use the center rectangle tool to draw a rectangle at the top of the 1.25-inch diameter center hole.

33. Use the Point tool to locate a sketch point at the bottom of the 1.25-inch hole.

34. Use the Smart Dimension tool to size the keyway as shown in the following figure between the point just placed and the top of the keyway.

35. Exit Sketch

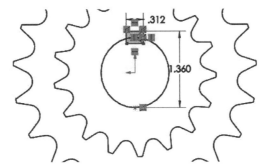

36.  With **Sketch4** highlighted, pick the Extruded Cut icon. For Direction 1 pick **Through All**. Pick the green checkmark.

37. Use the Chamfer tool to create the two welds. Select where the small sprocket meets the hub and where the larger sprocket meets the hub. Set the chamfer to 0.12 inches x 45°. Pick the green checkmark.

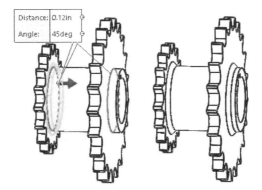

We are almost done; however, when we look at the original dual-toothed sprocket and our model we notice that the teeth are chamfered on the original sprocket. If we do this now, we will have to pick the edge of every tooth on the sprocket. If we chamfered the edge before we cut the teeth, we would only need to pick four edges. Let's back our model up to the point before we cut the first sprocket tooth and add a chamfer of 0.030 by 45 degrees.

38. Drag the blue line back up to just after Revolve1 to roll our model back.

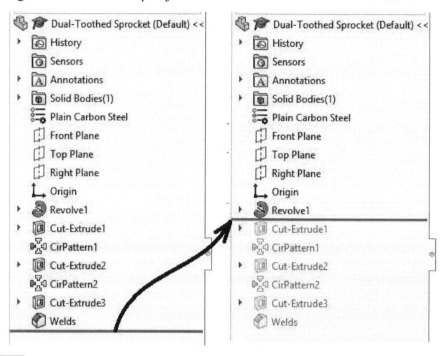

39. 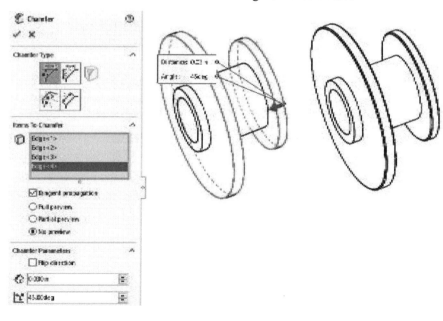 Select the Chamfer tool. Pick the four edges of the sprocket blank. Set the size of the chamfer to 0.03 inches x 45°. Pick the green checkmark.

40. Drag the blue line back down to the bottom of the list.

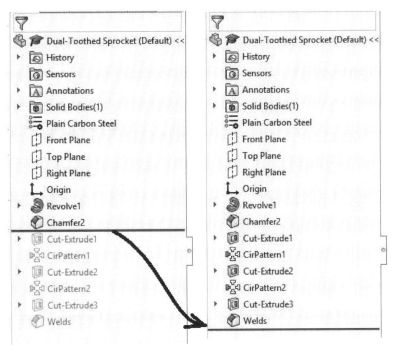

41. Right-click on **Material <not specified>** in the **Feature Manager Design Tree**, then select **Plain Carbon Steel** from the list.

42. Save the Part by picking the Save icon or at the top select **File> Save**.

43. Name the part file "**Dual-Toothed Sprocket**", then pick the **Save** button.

44. Select **File> Print...** (if asked to)

45. Select **File> Close** or type <Ctrl>+W to close the part and clear the window.

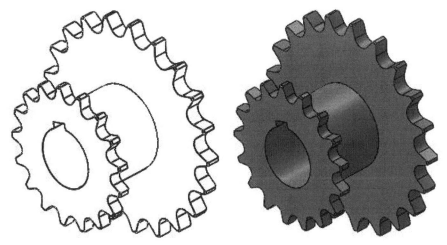

Patterned Features Problems

5-1 Design **"Bearing 232"** made from AISI 4340 annealed steel. The origin is on the centerline at the left end of the bearing. The four holes are evenly spaced on a 7.25-inch diameter bolt circle (BC). Note that the groove with the .25-inch diameter hole is .19 inches deep. **Design change:** Increase the 7.25-inch bolt circle to 7.50 inches. Note: The left side round is R.19 inches; all the rest of them are R.12 inches. Also, add a .12x45° chamfer to both ends of the center hole.

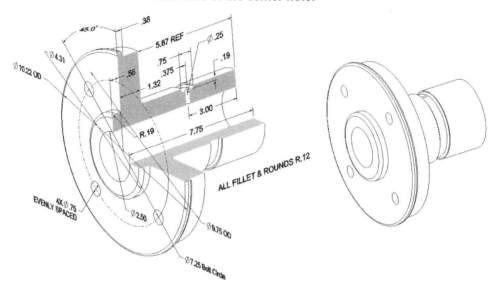

5-2 Design the **"Female Sleeve"** made from malleable cast iron. The origin is on the centerline at the front of the sleeve. The center hole is 18.0 inches in diameter. Add a 0.25-inch fillet where the 20-inch diameter meets the 26-inch diameter. The eighteen holes are evenly spaced on a 23.0-inch diameter bolt circle (BC). **Design change:** Increase the 1.25-inch diameter holes to 1.28 inches. All dimensions are in inches.

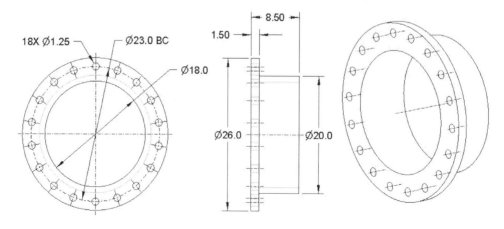

5-3   Design the **"Fixed Bearing Cup"** made from AISI 4130 steel annealed at 865° C. The origin is on the centerline at the back of the part. Add fillets and rounds to the 3D model last. Don't round the required sharp edges. The five holes are evenly spaced on a 4.25-inch diameter bolt circle (BC). The 40-degree cutout is symmetric about a vertical centerline that goes through the origin and aligns with the back surface of the bearing cup. See Figure below. Also, note that some dimensions are three decimal places while others are two decimal places.

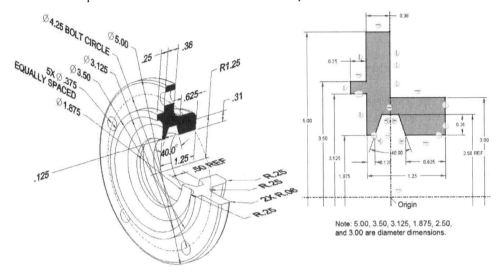

Note: 5.00, 3.50, 3.125, 1.875, 2.50, and 3.00 are diameter dimensions.

5-4   Design the straight-sided spline drive fitting for the **"Clutch Cone 15hp"** designed in problem 4.1 and add it to the clutch cone. The fitting has 10 splines evenly spaced as shown below. For a 10-toothed spline that will slide when not under load and with an outside diameter of "D," the inside diameter is 0.860*D, and the width of the cutout is 0.156*D. The allowable torque of the spline is $326*D^2L$ in.lb. A 15 horsepower engine at 3400 r.p.m. generates 278 in.lbs. of torque. The cone's spline will be 2.24 inches long with an average diameter of 0.930 inches; therefore, this spline can handle 630 in.lb. of torque. This provides a safety factor greater than 2.2. Add a 0.06-inch x 45° chamfer to both ends of the splined hole. Suggestion: make the hole 0.86 inches, then cut the spline teeth.

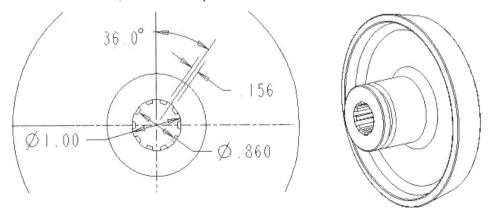

5-5 Design the **"End Plate"** made from plain carbon steel. The 2.00-inch diameter spot face (SF) is 0.03 inches deep.

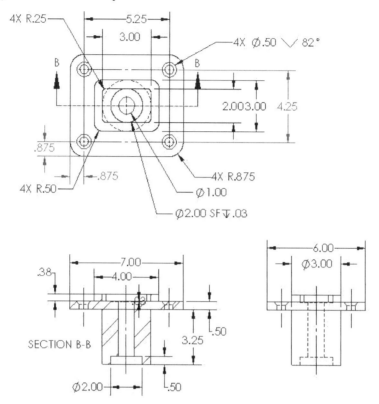

5-6 Design the **"Erector set Angle Link"** made from plain carbon steel. Suggested procedure. Create the cross-section of the link, then extrude it to 6.375 inches. Add a straight slot as shown to one end of the link, then create 13 incidents, 0.50 inches apart using the linear pattern tool. Use the circular pattern tool to rotate the pattern 90 degrees about the inner intersection of the two sides to create holes on the other side of the link.

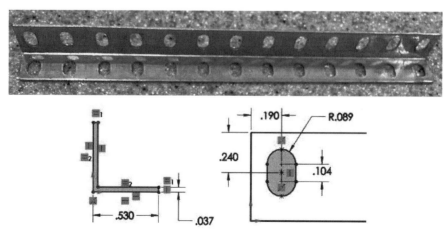

5-7   Design the **"Erector set Swivel Plate"** made from plain carbon steel. Suggested procedure. Create the cross-section of the link, then extrude it 3.00 inches about the Mid Plane. Create a 0.173-inch diameter hole Through All, then use the linear pattern tool to create 11 incidents 0.25 inches apart. Click on the Instances to Skip window, then pick the 2nd, 5th, 7th, and 10th holes. Do a similar procedure on the flat surface. Finally, use the circular pattern tool to create the circular hole pattern ½ inch from the center hole about the origin.

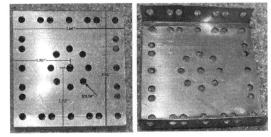

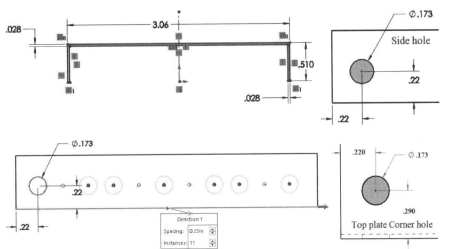

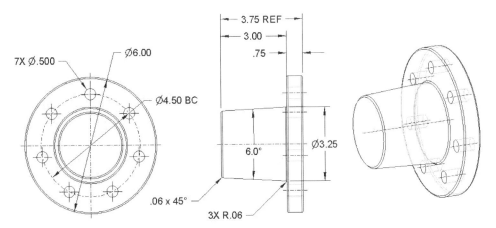

5-8   Design the **"Pipe Plug"** made from malleable cast iron. The origin is on the centerline where the taper touches the larger diameter. The seven holes are evenly spaced on a 4.50-inch diameter bolt circle (BC). **Design change:** increase the 0.500-inch diameter holes to 0.505 inches.

5-9   The **"Erector set Base Plate 1936"** from the A.C. Gilbert Company 1936 Erector
Set needs to be replaced so that the Worm Gear Drive shown below can be
assembled. The part is made from plain carbon steel.

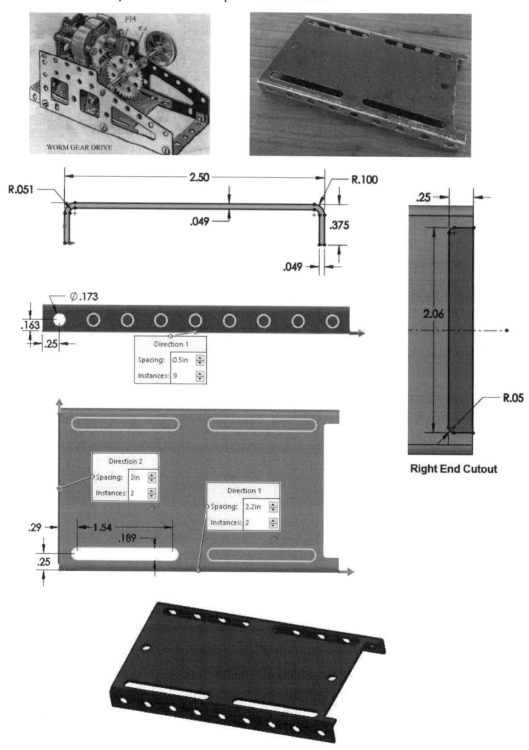

5-10   The **"Erector set Gear Box Base"** from the A.C. Gilbert Company 1936 Erector
        Set needs to be replaced so that the Worm Gear Drive shown in Problem 5-9 can
        be assembled. The part is made from plain carbon steel. All holes have a diameter
        of .173 inches. Typical spacing between holes is .50 inches.

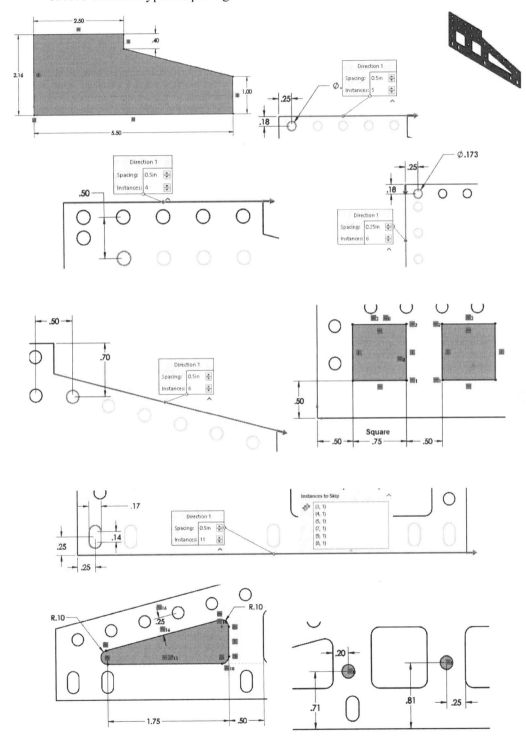

5-11   Design the **"Erector set Flat Plate 1x4"** from the A.C. Gilbert Company 1936 Erector Set made from plain carbon steel.

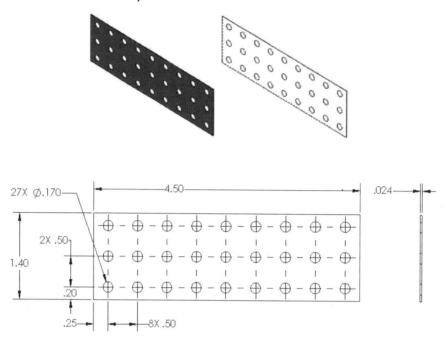

5-12   Design the **"Erector set 2.5-inch Dia Plate"** from the A.C. Gilbert Company 1961 Erector Set made from plain carbon steel. The eight holes in the two circular hole patterns are equally spaced.

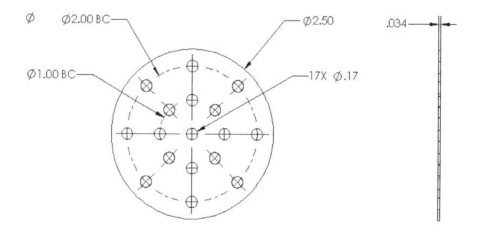

5-13 Design the **"End Cap"** made from plain carbon steel. Use the circular pattern tool to create the second and third holed tabs after creating the first tab. The three holes should be created using a 33/64-inch drill.

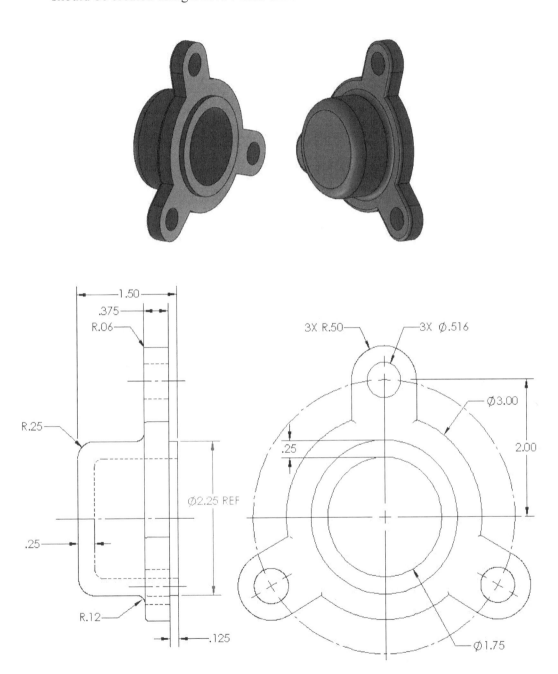

5-14 Design the **"Stanchion Support"** made from 1060 Alloy Steel. When creating the part, create one rib, then use the circular pattern tool to create the other three ribs. Do the same thing for the .531-inch diameter holes. Create one hole, then use the circular pattern tool to create the other three holes. Add a general note to break all sharp edges. The center hole is 1.75 inches deep.

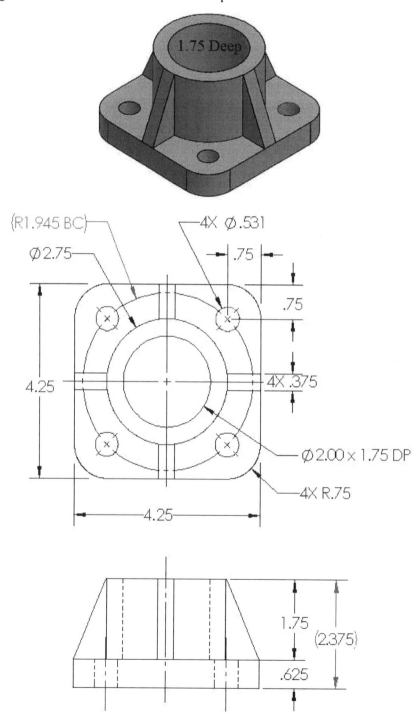

Chapter 6.   Dimensioning

Introduction

Before the 1800s an inch was defined as the width of a man's thumb, and a foot was simply the length of a man's foot. (In 1793 France adopted the metric system based on the meter.) In 1824 the English Parliament defined the length of a yard. A foot was one-third of a yard, and an inch was one-twelfth of a foot. From these specifications, graduated rulers were developed. Until the twentieth century, common fractions were adequate for dimensions, but as machines became more complicated, more accurate specifications were required. It became necessary to express dimensions as decimal numbers.

In addition to a complete shape description of a part, an engineering drawing must give a complete size description. In the early 1900s, the design and production functions were located in one factory. In some cases, the complete part creation was carried out by the same worker. Design drawings were nothing more than assembly drawings without dimensions. The worker used a scale directly on the drawing to obtain the required dimensions. It was up to the worker to make sure the parts fit together properly. If a question came up, the worker would consult the designer who was always nearby. Under these manufacturing conditions, drawings didn't need to carry detailed dimensions or notes.

The modern methods of size description came into existence with the need for interchangeable parts. Detailed drawings must be dimensioned so that a worker can produce mating parts that can be assembled in a separate factory or used as replacement parts by the customer. The need for precision manufacturing and the controlling of sizes for interchangeability shifted from the machinist to the designing engineer. The worker no longer used his judgment in engineering matters but rather executed the instructions given on the detailed drawings. Thus, design engineers needed to become familiar with the materials and the processes of the shop.

The drawings must show the part in its completed form and contain all information necessary to make it. Therefore, when dimensioning a detailed drawing, the engineer must keep in mind the finished part, the shop processes required, and the design intent of the part. Dimensions need to be given that are required by the worker when making the part. There is no reason to provide dimensions to points or surfaces that are not accessible to the worker.

Dimensions must not be duplicated on a drawing. Only the dimensions needed to produce and inspect the part, as intended by the design engineer, should be shown. The beginner often makes the mistake of giving the dimensions he/she used to make the computer-generated part model or the drawing. These are not necessarily the dimensions required by the shop and should not be on the engineering drawing.

Drawings must be made to scale with the scale listed in the title block.

Learn to Dimension Properly

Dimensions are given in the form of linear distances, angles, and notes. The engineer must learn the skills of dimensioning; that is the type of lines to use, the spacing between dimensions, proper note creation, etc. A sample detailed part drawing is shown below without the border and title block.

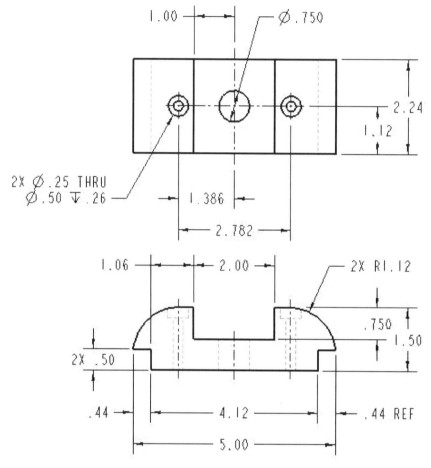

The beginner must learn the rules for placing dimensions on a detailed drawing. The suggestions to follow provide a logical arrangement for maximum legibility. The part's function is considered first; the shop processes second. The proper procedure is to dimension for design intent, and then review the dimensions to see if any improvements in clarity can be made without affecting design intent.

A dimension line is a dark, solid line terminated by arrowheads, which indicates the direction and extent of a dimension. On detailed part drawings, the dimension line is broken near the middle to provide an open space for the dimensional value.

The dimension lines closest to the part should be spaced at least 1/2 inch away, as shown below. All other parallel dimension lines should be at least 3/8 inches apart. The spacing of dimension lines should be uniform throughout.

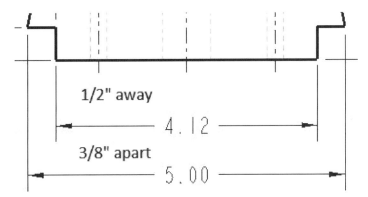

An extension line is a dark, solid line that "extends" from a point near the feature to the corresponding dimension. The dimension line meets the two extension lines at a right angle. A gap of about 1/16 of an inch should be left between the extension line and the part feature. The extension line should extend about 1/16 of an inch beyond the outermost dimension line. These are the defaults for SOLIDWORKS.

A centerline is a dark line composed of alternate long and short dashes and is used to represent the axes of symmetrical parts or to denote centers. As shown below, do not create a gap when the centerline crosses the part outline. A centerline must end with a long dash. These are the defaults for SOLIDWORKS.

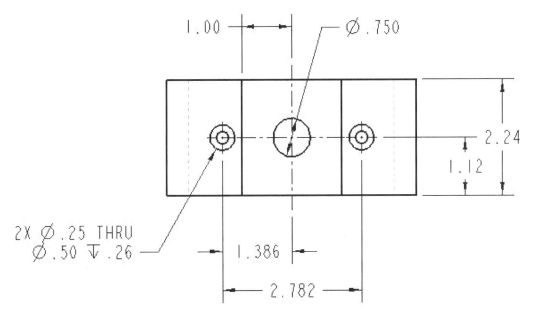

An example of the placement of dimension lines and extension lines is shown above. The shorter dimensions are closest to the part outline. Dimension lines must not cross extension lines. This would happen if the shorter dimensions were placed outside the longer dimensions. Do not do this. Note that it is acceptable for extension lines to cross each other. Extension lines should never be shortened. A dimension line must never coincide with or form a continuation of any part outline.

Do not allow dimension lines to cross. Dimensions should be lined up and grouped as much as possible. In general, never dimension to hidden lines. In some cases, extension lines and centerlines may cross visible lines of the part. When this occurs, a gap must not be created.

A leader is a thin, solid line leading from a note or dimension to the specified feature. It is terminated by an arrowhead touching the part feature to which the note applies. An arrowhead must stop on a line, such as the edge of a hole. A leader should be an inclined straight line, except for its short horizontal shoulder extending from the mid-height of the beginning or end of the note.

A leader to a circle must be radial so that if extended it would pass through the center of the circle. A drawing presents a more pleasing appearance if leaders near each other are drawn parallel. Leaders should cross as few lines as possible, and should never cross each other. Leader lines should not be drawn parallel to nearby lines of the part drawing, allowed to pass through the corner of a part, be drawn excessively long, or drawn nearly horizontally or vertically. A leader line should be drawn at an angle between 20° and 70°.

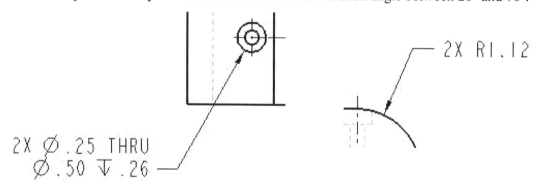

Fractional and Decimal Dimensions

In the early 1900s, the worker scaled the design drawing without dimensions to obtain any necessary values, and he was responsible for making the parts fit together properly. When blueprinting became common, workers in separate factories would use the same drawings to make similar parts, thus it was necessary to dimension the drawings in common fractions and inches. The smallest dimension a worker could measure directly was 1/64 of an inch. This became the smallest division on the machinist's scale. When close fits were required, the drawing would carry a note, such as "running fit," or "force fit," and the worker would make a small adjustment to the size. Machinists were very skilled. Hand-built machines were often beautiful examples of precision workmanship.

Today there are many types of products where common fractional inches are used because extreme accuracy is not necessary. A boxcar drawing, in which the structure is very large, does not require great accuracy. Also, there are many parts where the ordinary machinist's scale is accurate enough.

As the manufacturing industry progressed, there became a greater demand for more accurate specifications of the important functional dimensions. Since it was hard to use smaller fractions, such as 1/128 of an inch, it became common practice to give dimensions as decimal values, such as 3.62, 1.375, or 2.1625. However, many relatively unimportant machine dimensions, standard nominal sizes of materials, drilled holes, standard threads, keyways, etc., can be expressed as whole numbers with common fractions.

Thus, a given detailed drawing could be dimensioned with whole numbers and common fractions, with decimals, or a combination of the two. The latter is common, especially with machined castings and forgings, where the rough dimensions need not be closer than ±1/64 of an inch. However, because of computer-aided design and the requirements for accuracy, the standard practice has moved toward the adoption of the decimal system only. In addition, the decimal system is compatible with numerically-controlled machines.

The decimal system, based on the inch, has many of the advantages of the universal metric system and is compatible with most measuring devices and machine tools. American manufacturers have found that the decimal inch system, rather than the metric system, works without the need to scrap their inch-type measuring devices. However, because products are now exported worldwide, the metric system has become more common for these products.

In 1932, the Ford Motor Company adopted a complete decimal system. The shop scale chosen was divided on one edge into inches and tenths of an inch, and on the other edge into inches, tenths, and fiftieths of an inch. Thus, the smallest division was one-fiftieth or .02 inches; two divisions were .04 inches, etc. When it was necessary to halve a value, the result would still be a two-place decimal. The automotive and aircraft industry used this system for many years.

Two-place decimals are used when a tolerance of ±.01 inch or greater is allowed. Three or more decimal places are used for tolerance limits less than ±.01 inches. In the standard two-place decimal system, the second place should be an even digit (for example, .02, .04, and .06 are preferred over .01, .03, or .05). When a dimension is divided by two, such as in determining the radius from a diameter, the result is still a two-place decimal. Odd two-place decimals are allowed when required for design purposes.

In this system, common fractions may be used to indicate nominal sizes of materials, drilled holes, punched holes, threads, keyways, and other standard features. For example, 1/4-20 UNC-2A; 3/4 DRILL; or STOCK 1½ x 1½. If desired, decimals may be used for everything, including standard nominal sizes, such as .25-20 UNC-2B, or .75 DRILL.

When a decimal value is to be rounded to a lesser number of decimal places, do this:

1. When the digit beyond the last kept digit is less than 5, the last digit kept does not change. Example: 1.1624 is rounded to three decimal places, or 1.162.

2. When the digit beyond the last kept digit is more than 5, the last kept digit is increased by 1. Example: 2.2768 is rounded to three decimal places, or 2.277.

3. When the digit beyond the last kept digit is exactly 5 with only zeros following if the kept digit is even, it is unchanged; if it is odd increase it by 1. Example: 3.565 becomes 3.56 when rounded to two decimal places, and 3.375 rounds to 3.38.

For the decimal equivalents of common fractions, see the following Table of Fractions and Decimal Equivalence.

Table 6-1 Fractions and Decimal Equivalence			
Fraction	**2-decimals**	**3-decimals**	**4-decimals**
1/64	0.02	0.016	0.0156
1/32	0.03	0.031	0.0313
3/64	0.05	0.047	0.0469
1/16	0.06	0.063	0.0625
5/64	0.08	0.078	0.0781
3/32	0.09	0.094	0.0938
7/64	0.11	0.109	0.1094
1/8	0.13	0.125	0.1250
9/64	0.14	0.141	0.1406
5/32	0.16	0.156	0.1563
11/64	0.17	0.172	0.1719
3/16	0.19	0.188	0.1875
13/64	0.20	0.203	0.2031
7/32	0.22	0.219	0.2188
15/64	0.23	0.234	0.2344
1/4	0.25	0.250	0.2500
17/64	0.27	0.266	0.2656
9/32	0.28	0.281	0.2813
19/64	0.30	0.297	0.2969
5/16	0.31	0.313	0.3125
21/64	0.33	0.328	0.3281
11/32	0.34	0.344	0.3438
23/64	0.36	0.359	0.3594
3/8	0.38	0.375	0.3750
25/64	0.39	0.391	0.3906
13/32	0.41	0.406	0.4063
27/64	0.42	0.422	0.4219
7/16	0.44	0.438	0.4375
29/64	0.45	0.453	0.4531
15/32	0.47	0.469	0.4688

Table 6-1 Fractions and Decimal Equivalence (Continued)			
Fraction	2-decimals	3-decimals	4-decimals
31/64	0.48	0.484	0.4844
1/2	0.50	0.500	0.5000
33/64	0.52	0.516	0.5156
17/32	0.53	0.531	0.5313
35/64	0.55	0.547	0.5469
9/16	0.56	0.563	0.5625
37/64	0.58	0.578	0.5781
19/32	0.59	0.594	0.5938
39/64	0.61	0.609	0.6094
5/8	0.63	0.625	0.6250
41/64	0.64	0.641	0.6406
21/32	0.66	0.656	0.6563
43/64	0.67	0.672	0.6719
11/16	0.69	0.688	0.6875
45/64	0.70	0.703	0.7031
23/32	0.72	0.719	0.7188
47/64	0.73	0.734	0.7344
3/4	0.75	0.750	0.7500
49/64	0.77	0.766	0.7656
25/32	0.78	0.781	0.7813
51/64	0.80	0.797	0.7969
13/16	0.81	0.813	0.8125
53/64	0.83	0.828	0.8281
27/32	0.84	0.844	0.8438
55/64	0.86	0.859	0.8594
7/8	0.88	0.875	0.8750
57/64	0.89	0.891	0.8906
29/32	0.91	0.906	0.9063
59/64	0.92	0.922	0.9219
15/16	0.94	0.938	0.9375
61/64	0.95	0.953	0.9531
31/32	0.97	0.969	0.9688
63/64	0.98	0.984	0.9844
1	1.00	1.000	1.0000

The reading direction of dimensions for the two systems has been approved by the American Standards Association or ASA. In the unidirectional system, all dimensions and notes are printed horizontally and are read from the bottom of the drawing. This is the preferred system. In the aligned system, all dimensions are aligned with the dimension lines so that they may be read from the bottom or the right side of the drawing. Notes are always positioned horizontally in both systems.

Standard Sizes Preferred

Dimensions should be given, wherever possible, to make use of available materials, tools, parts, and gages. The dimensions for many commonly used machine elements, such as bolts, nails, keys, tapers, wire, pipes, sheet metal, chains, belts, and pins have been standardized. The design engineer must obtain these standard sizes from published handbooks, from the American Standards Association, or the manufacturers' catalogs.

Detailed drawings of these standard parts are not made unless the parts are to be modified. They are drawn conventionally on the assembly drawing and are listed in the bill of materials. Common fractions are generally used to indicate the nominal sizes of standard parts. If the decimal system is used exclusively, all sizes are expressed in decimals; for example, .25 DRILL instead of ¼ DRILL.

Dimensioning Angles

Angles are dimensioned by means of the coordinate dimensions of the two legs of the right triangle, or by means of a linear dimension and an angle in degrees. Coordinate dimensions are suitable for work requiring a high degree of accuracy. Variations of the angle are hard to control because the amount of variation increases with the distance from the vertex. Methods of indicating various angles are shown below.

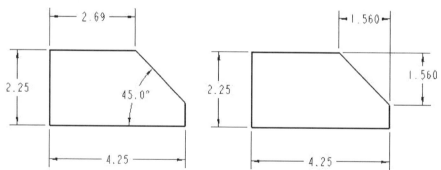

When degrees alone are indicated, the symbol ° is used. When minutes alone are given, the value should be preceded by 0°, for example, 0° 36'. However, it is preferred that an angle be given in degrees with one decimal place, such as 38.6°. Note that there is no gap where the extension lines cross in the right view.

Dimensioning Arcs

A circular arc is dimensioned in the view where its true shape is shown by giving the numerical value of its radius, preceded by the letter R. The number of arcs with the same radius is indicated in front of the R-value. The center of the arc may be indicated by a small cross. When there is enough room, both the numerical value and the arrowhead may be placed inside the arc. For smaller arcs, the arrowhead may be left inside and the numerical value may be placed outside. For small arcs, both the arrowhead and the numerical value are placed outside the arc.

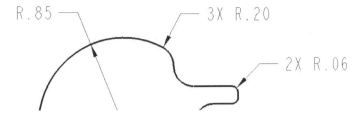

Fillets and Rounds

Individual fillets and rounds are dimensioned as arcs. If there are only a few arcs and they are the same size, specifying one typical radius is sufficient. However, fillets and rounds are often quite numerous on a drawing and most of them are likely to be some standard size, such as R.25. In these cases, it is customary to provide a general note on the drawing to cover all uniformly-sized fillets and rounds, thus: "FILLETS R.25 AND ROUNDS R.12 UNLESS OTHERWISE SPECIFIED" or simply "ALL FILLETS AND ROUNDS R.25."

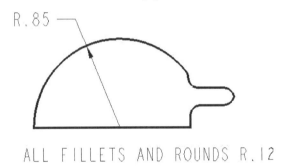

ALL FILLETS AND ROUNDS R.12

Finish Marks

A finish mark is used to indicate that a surface is to be machined or finished, such as on a rough casting or forging. To the die maker, a finish mark means that he must allow extra material for the rough workpiece. The number by the symbol is the surface roughness in micro-inches. A triangular symbol means that machining is mandatory. A surface finish of 8 micro-inches could be done by grinding or honing the surface. The finishing mark with the circle in the V indicates machining prohibited with a maximum surface roughness of 125 micro-inches.

The following table lists the typical shop operations and their expected surface roughness when machining metal in both average micrometers and average micro-inches. To convert micro-inches to micrometers, divide micro-inches by 40.

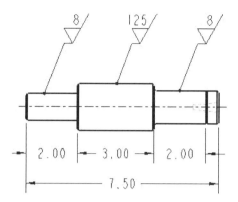

A new system of roughness grades has come into being recently. It should be used on drawings for international suppliers and for new designs. It grades the surface roughness with a value from 1 to 12 with 12 being the roughest surface. N12 is equivalent to 2000 μ-inches or 50 μ meters; N11 is equivalent to 1000 μ-inches or 25 μ meters; N10 is equivalent to 500 μ-inches or 12.5 μ meters, etc., down to N1 being equivalent to 1 μ-inch or .025 μ meters.

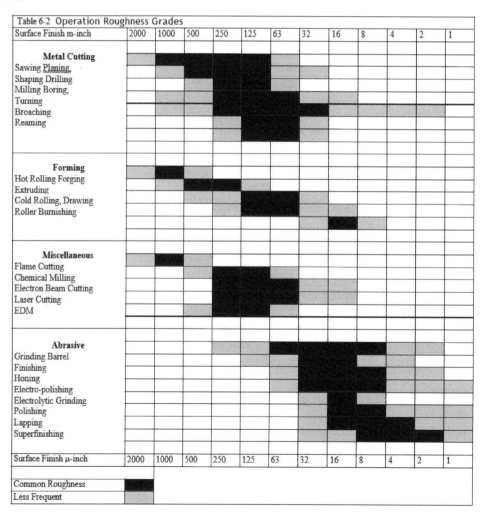

Table 6-2 Operation Roughness Grades

Surface Finish m-inch	2000	1000	500	250	125	63	32	16	8	4	2	1

Metal Cutting — Sawing Planing, Shaping Drilling, Milling Boring, Turning, Broaching, Reaming

Forming — Hot Rolling Forging, Extruding, Cold Rolling, Drawing, Roller Burnishing

Miscellaneous — Flame Cutting, Chemical Milling, Electron Beam Cutting, Laser Cutting, EDM

Abrasive — Grinding Barrel, Finishing, Honing, Electro-polishing, Electrolytic Grinding, Polishing, Lapping, Superfinishing

Surface Finish μ-inch	2000	1000	500	250	125	63	32	16	8	4	2	1

Common Roughness	
Less Frequent	

Dimensions and Part Views

Dimensions should not be placed on a view unless clearness is promoted. The ideal dimensioning scheme is shown below where all dimensions are placed outside the part view. Compare this to the poor practice shown to its right. This is not to say that a dimension should never be placed on a part.

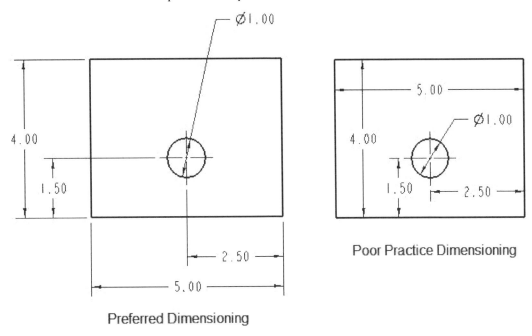

Preferred Dimensioning

Poor Practice Dimensioning

When a dimension must be placed in a sectioned area, as shown below, provide a blank space in the sectioned area for the dimensional value.

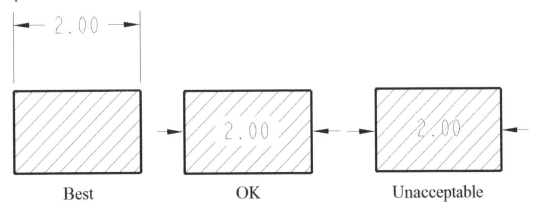

Best   OK   Unacceptable

Contour Dimensioning

Views are drawn to describe the shapes of the various features of the part, and the dimensions are given to define the exact sizes and locations of those features. It follows that a dimension should be given where the shape is shown true, that is, in the view where the contour is obvious such as in Figure (a) on the left. The incorrect placement of the dimensions is shown in Figure (b) on the right.

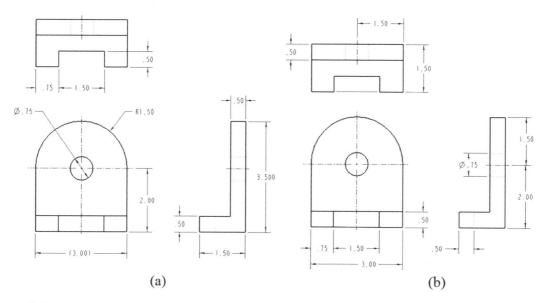

(a)   (b)

Individual dimensions should be attached directly to the contours that show their shape. This will prevent the attachment of dimensions to hidden lines. Also, this will prevent the attachment of a dimension to a visible line where the meaning is not clear.

Although the placement of notes for holes follows the contour principle, the diameter for a solid cylindrical shape is given in the rectangular view so it can be located near the dimension for the cylinder's length.

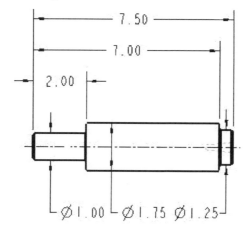

Geometric Shapes

Mechanical parts are usually composed of simple geometric shapes, such as prisms, cylinders, cones, or spheres. They may be exterior (positive) or interior (negative) shapes. For example, a step shaft is made up of positive cylinders, while a counter-bored hole is made up of negative cylinders.

These features result directly from the need to keep shapes as simple as possible, and from the requirements of the basic shop operations. Features with a plane surface can be produced by planing, shaping, or milling, while features having a cylindrical, conical, or spherical surface can be produced by turning, drilling, reaming, counter boring, boring, or countersinking.

The dimensioning of engineering parts begins by giving the dimensions showing the sizes of the simple geometric shapes, then giving the dimensions locating these features with respect to each other. The former is called "size dimensions" and the latter is called "location dimensions." This method of geometric analysis is very helpful when dimensioning any part, but must be altered when there is a conflict with the function of the part.

A 2-view drawing of a part is shown below. The geometric shape is dimensioned with size dimensions, and then the features are located with respect to each other using location dimensions. Note that a location dimension locates a three-dimensional feature and not a surface; otherwise, all dimensions would be classified as location dimensions.

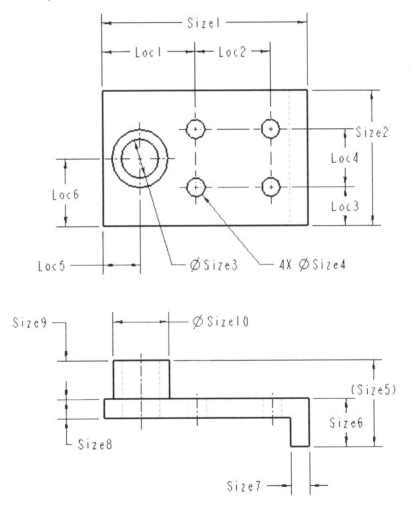

Dimensioning Prisms

The right rectangular prism is the most common geometric shape. The front and top views are dimensioned as shown below. The height and width are given in the front view and the depth is in the top view. A dimension between the views applies to both views and should not be placed elsewhere without good reason. The dimensions may be placed on the right side or the left side of the views.

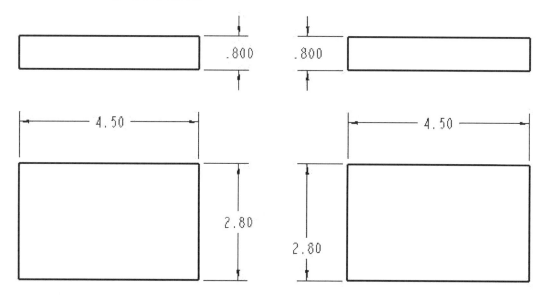

The application of size dimensions to a machine part composed entirely of rectangular prisms is shown below. The 3.40-inch dimension is listed as a reference dimension since it is not necessary.

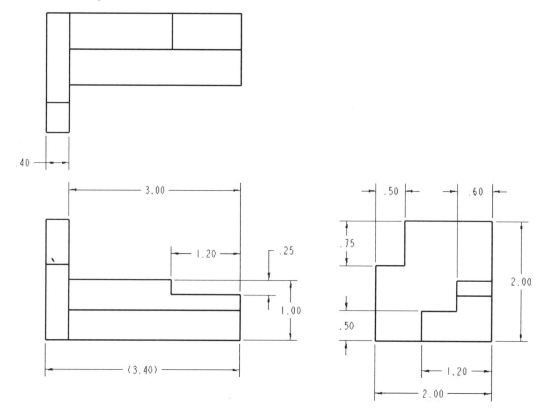

Dimensioning Cylinders

The right circular cylinder is the next most common geometric shape and is commonly seen as a shaft or a hole. The general method of dimensioning an external cylinder is to give both its diameter and its length in the rectangular view. If the cylinder is drawn in a vertical position, the length or height of the cylinder may be given to its right or left.

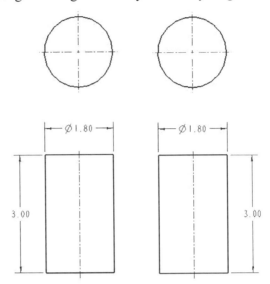

If a cylinder is drawn in its horizontal position, its length may be given above the rectangular view or below it. An application showing the dimensioning of cylindrical shapes is shown below. The radius of a cylinder must never be given since measuring tools are designed to check diameters.

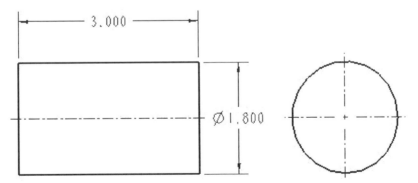

Dimensioning Holes

Holes that are to be drilled, bored, reamed, or punched are specified using standard notes specifying the diameter and depth. The order of items in a note should correspond to the order of the procedure in the shop when producing the hole. The note may include the shop processes. Two or more holes should be dimensioned using a single note that includes the number of holes preceding the first line of the note with the leader pointing to one of the holes.

As shown below, the leader of the note should point to the circular view of the hole. It can point to the rectangular view only when clearness is promoted such as in a sectioned view. When the circular view of the hole has two or more concentric circles, as with a counterbored, countersunk, or tapped hole, the arrowhead should touch the outer circle.

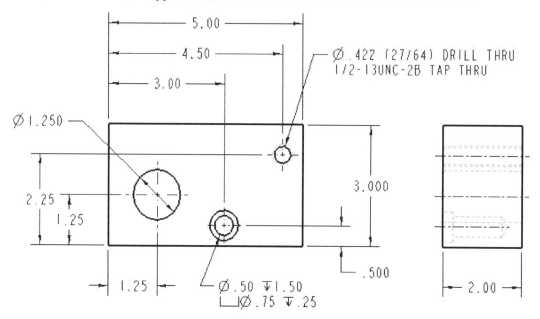

The use of decimal fractions instead of common fractions to designate drill sizes has gained wide acceptance. For numbered or letter-size drills, it is preferred that the decimal size be given as well. For example, a numbered drill would be shown as #20 (ø.161) DRILL or a letter size drill as "K" (ø.281) DRILL.

On part drawings where the parts are to be produced in large quantities, dimensions, and notes may be given without specification to the shop process. Even though the shop operations are omitted, the tolerances will dictate the shop processes needed.

Dimensioning Round-End Shapes

The method for dimensioning round-end shapes depends upon the degree of accuracy required. When precision is not necessary, the dimensioning method used is the one that is convenient for the shop.

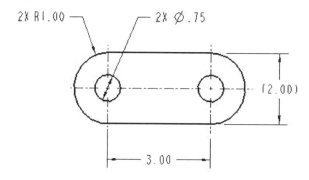

The link shown above is to be cut from sheet metal, thus it should be dimensioned as it would be laid out in the shop, that is, by giving the hole's center-to-center distance and the radii at the ends. Only one radius dimension is necessary. If an overall width dimension is shown, it must be shown as a reference dimension. The same would be true if an overall length was given.

The pad on a casting with a milled slot shown below is dimensioned from center to center for the convenience of both the patternmaker and the machinist. An additional reason for the center-to-center distance is that it gives the total travel of the mill cutter, which can be easily controlled. The width dimension indicates the diameter of the mill cutter; hence, it would be inappropriate to give the radius of the machined slot.

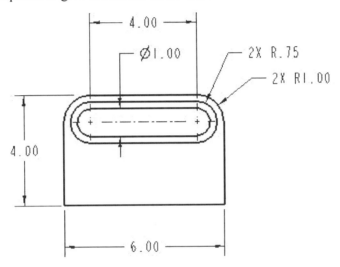

When accuracy is required, the dimensioning method shown below is recommended. Overall lengths of rounded end shapes are given in each case, and radii are indicated, but without a specific value or listed as REF as indicated by parentheses.

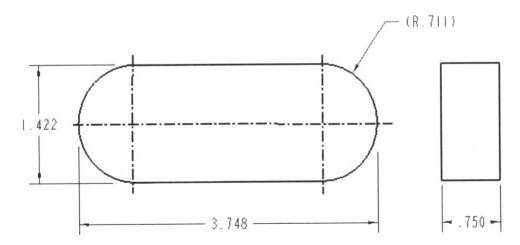

Dimensioning Tapers

A taper is a conical surface on a shaft or a hole. The method of dimensioning a taper is to provide the amount of taper per unit length in a note: "TAPER .025 INCH per INCH," and then give the diameter at one end, along with the taper length.

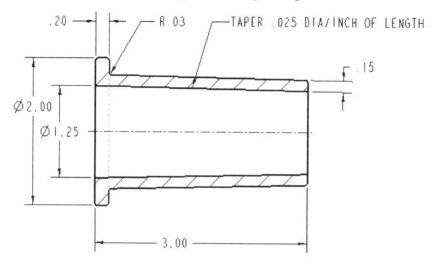

Standard machine tapers are used on machine spindles, shanks of tools, and pins. They are dimensioned on a drawing by dimensioning the larger diameter and the taper size in a note, such as "NO. 4 AMERICAN STANDARD TAPER (.0519 inch per inch)."

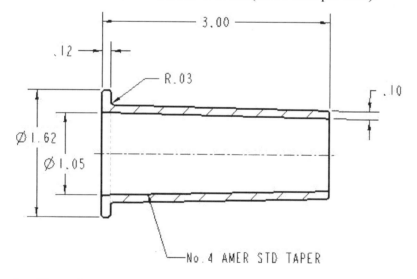

Dimensioning Threads

Local notes are used to specify thread dimensions. For a tapped hole, the note should be attached to the circular view of the hole. The note is placed in the longitudinal view for external threads where the threads are easily recognized. The following table is the ANSI Standard V-thread tap size table for Unified National Coarse and Fine Threads.

Tap Drill Sizes for Fine and Coarse Threads				
Diameter	Fine Threads	Tap Drill Size	Coarse Threads	Tap Drill Size
#0	80	3/64	–	–
#1	72	No. 53	64	No. 53
#2	64	No. 50	56	No. 50
#3	56	No. 45	48	No. 47
#4	48	No. 42	40	No. 43
#5	44	No. 37	40	No. 38
#6	40	No. 33	32	No. 36
#8	36	No. 29	32	No. 29
#10	32	No. 21	24	No. 25
#12	28	No. 14	24	No. 16
1/4	28	No. 3	20	No. 7
5/16	24	I	18	F
3/8	24	Q	16	5/16
7/16	20	25/64	14	U
1/2	20	29/64	13	27/64
9/16	18	33/64	12	31/64
5/8	18	37/64	11	17/32
3/4	16	11/16	10	21/32
7/8	14	13/16	9	49/64
1	14	59/64	8	7/8

Dimensioning Chamfers

A chamfer is a beveled or sloping edge. When the angle is not 45°, it is dimensioned by giving the length of one leg and the angle, as shown below. A 45° chamfer is usually dimensioned using a note without the word CHAMFER or CHAM.

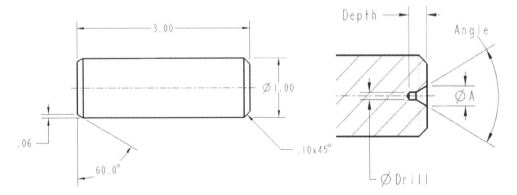

Shaft Centers

Shaft centers are required on shafts, spindles, and other conical parts for turning, grinding, or other rotational operations. Such a center may be dimensioned, as shown above. Normally the shaft centers are produced by a combined drill and countersink. The angle is typically 60°. The following Table shows the recommended shaft center sizes for given shaft diameters.

Shaft Center Sizes				
Shaft Diameter (inch)	Up to (inch)	A (inch)	Drill (inch)	Depth (inch)
0.1875	0.2499	0.08	0.047	0.090
0.2500	0.3749	0.09	0.047	0.103
0.3750	0.5124	0.13	0.167	0.042
0.5125	0.8124	0.19	0.078	0.188
0.8125	1.1249	0.25	0.094	0.229
1.1250	1.4999	0.31	0.156	0.292
1.5000	1.9999	0.38	0.094	0.400
2.0000	2.9999	0.41	0.219	0.355
3.0000	3.9999	0.50	0.219	0.462
4.0000	4.9999	0.56	0.219	0.516

Dimensioning Keyways

The method for dimensioning keyways for stock keys is shown below. A dimension is used to center the keyway on the shaft or collar. The preferred method of dimensioning the depth of a keyway is to give the dimension from the top or bottom of the keyway to the opposite side of the shaft or hole. The method of computing such a dimension for stock keys is shown below.

$$H = \frac{D - A + \sqrt{D^2 - A^2}}{2} = \frac{1.25 - 0.125 + \sqrt{1.25^2 - 0.125^2}}{2} = 1.185 \; inches$$

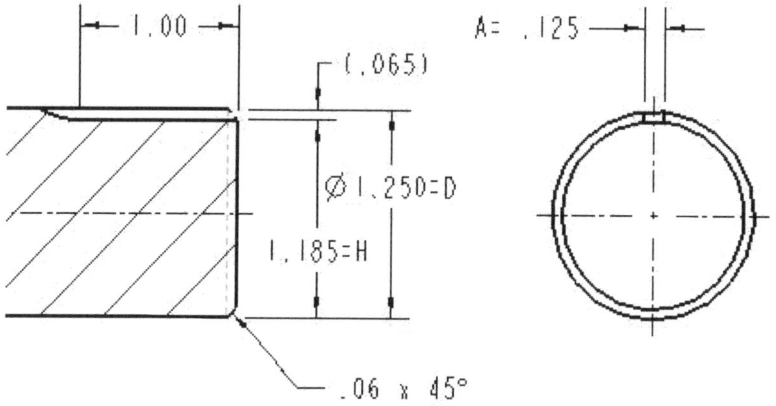

The method for dimensioning keyways for Woodruff keys is shown below. Values for the dimensions of Woodruff keys in a shaft can be found in the machinist's handbook.

Reference dimensions are shown below because they are listed in the handbook, and are not normally specified on the drawing.

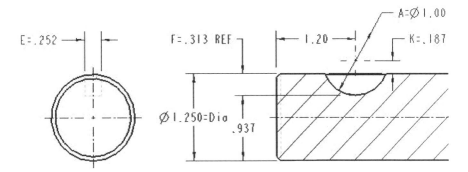

Cylindrical or conical shapes are located by their centerlines. Location dimensions for a hole are given in the circular view of the hole, as seen below.

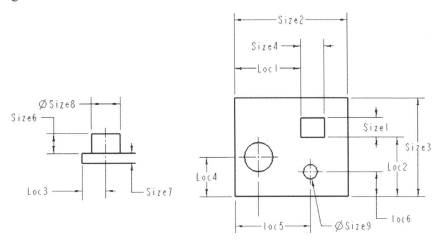

Location dimensions should come from finished surfaces wherever possible because rough castings and forgings vary in size, and unfinished surfaces cannot be relied upon for accurate measurements. The starting dimension, used in locating the first machined surface on a rough casting, will come from a rough surface or the center or the centerline of the rough part.

Location dimensions should reference a finished surface as a datum plane, or a centerline as a datum axis. When several cylindrical surfaces have the same centerline, do not locate them with respect to each other.

Holes equally spaced around a common center may be dimensioned by giving the bolt circle diameter and specifying "equally spaced" in the note if not obvious. Holes unequally spaced are located using the bolt circle diameter plus the angular measurement reference from one of the datum planes.

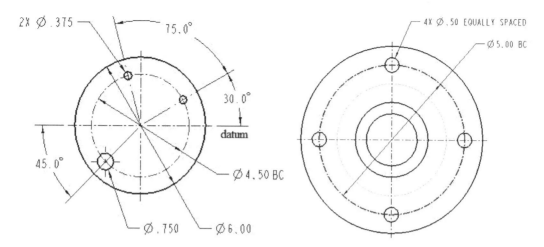

Where greater accuracy is required, coordinate dimensions should be given as shown below. In this case, the diameter of the bolt circle is marked REF to indicate that it is to be used only as a reference dimension. Reference dimensions are given for information purposes only. They are not intended to be measured and do not govern the shop processes. They represent calculated dimensions and are often useful in showing design intent.

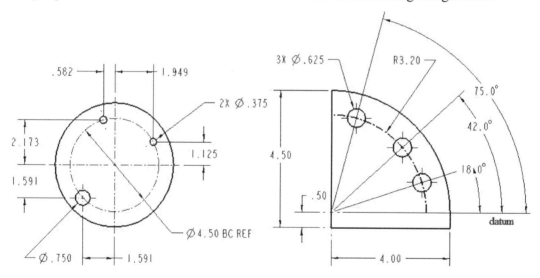

When several non-precision holes are located on a common arc, they are dimensioned by giving the radius of the arc and the angular measurements from a datum plane. See the figure above.

In the following figure, the three holes are on a common horizontal centerline. One dimension locates the right hole from the center; the other dimension gives the distance between the two small holes. Note the omission of the dimension from the center to the left hole occurs because the distance between the two small holes is the design intent. If the relationship between the center hole and each of the small holes was the design intent, then show it and mark the overall dimension between the two small holes as REF.

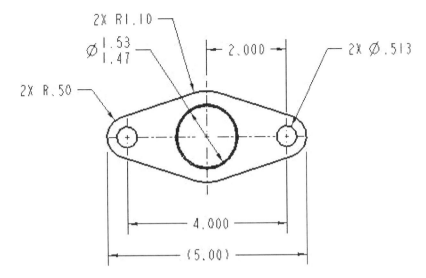

Another example of locating holes using linear dimensions is shown below. In this case, one measurement is made at an angle to the coordinate dimensions because of the design intent between the holes.

The holes are located from two baselines or datums (left side and bottom). When all holes are located from a common datum, the sequence of measuring and machining operations is controlled, overall tolerance accumulations are avoided, and the design intent of the finished part is assured. The datum surfaces selected must be more accurate than any measurement made from them and must be accessible during creation and setup to facilitate tool and fixture design. Thus, it may be necessary to specify the accuracy of the datum surfaces in terms of straightness, roundness, and flatness, thus Geometric Dimensioning and Tolerancing are required.

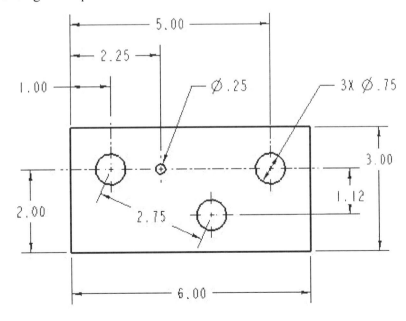

Modified Purchased Parts

In many machines, some parts are purchased and used as is. Sometimes the purchased parts need to be modified. For example, an assembly needs a C-channel of a given length. In this case, the detailed drawing shows only the dimensions needed to modify the purchased part. If other dimensions are shown below, they are listed as reference dimensions as indicated by the parentheses.

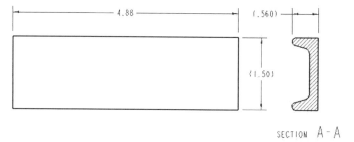

SECTION A-A

Mating Dimensions

When dimensioning a single part, its relationship to its mating parts must be taken into account. For example, a Block fits into the slot of its Base. The dimensions common to both parts are mating dimensions.

These mating dimensions should be given on the detailed drawings in their corresponding locations. The other dimensions are not mating dimensions since they do not control the mating of the two parts. The actual values of the two corresponding mating dimensions may be different. For example, the width of the slot may be dimensioned several thousandths of an inch larger than the width of the block, but these are the mating dimensions based on a basic width. The mating dimensions need to be identified so that they can be specified in the corresponding locations on the two parts, and so that they can be given with the degree of accuracy necessary for the proper fitting of the two parts.

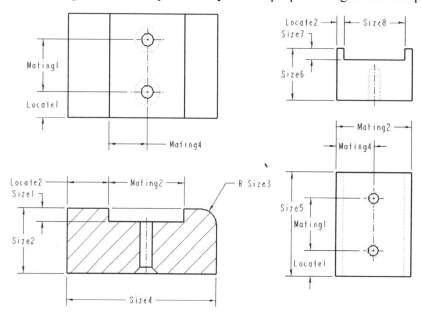

Notes

Sometimes it is necessary to supplement the direct dimensions with notes. The notes must be brief and carefully worded so as to have only one interpretation. Notes should not be created in crowded areas on the drawing sheet. Avoid placing notes between views. Notes should not be placed closer to a different view as to suggest the note applies to the wrong view. Note leaders should be short and cross as few lines as possible. They should never run through the corner of a part or any specific points or intersections on the part view.

Notes are classified as general notes when they apply to an entire drawing. They are referred to as local notes when they apply to specific features.

General notes should be placed in the lower right-hand corner of the drawing, above or to the left of the title block, or in a central position below the view to which they apply. Some general notes are "FINISH ALL OVER", "BREAK ALL SHARP EDGES TO R .06" or "DRAFT ANGLES @ 3° UNLESS OTHERWISE SPECIFIED." On detailed drawings, the title block should show some general notes on material and implied tolerances.

Local notes as shown below apply to specific features and are connected by a leader line to the point where the operation is performed. Some local notes are "4X 0.25 DRILL" or ".06 X 45° CHAM." The leader must be attached to the front of the first word or just after the last word.

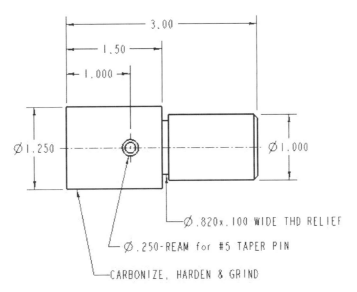

In general, leaders and local notes should not be located on the drawing until the dimensions are finalized. If notes are placed first, they typically have to be moved when locating the dimensions.

Certain commonly used abbreviations may be used in notes, such as THD or MAX. All abbreviations should conform to the American Standard Association.

Dimensioning for Numerical Control

In general, the basic dimensioning practices listed here are compatible with the data requirements for numerically controlled machines. However, to make the best use of this type of production, the designer must consult the manufacturing machine manuals before making the drawings. Certain considerations should be noted:

1. Coordinate dimensioning is required from three mutually perpendicular references or datum planes. These must be clearly identified on the drawing.
2. The reference planes may be located on or off the part, but preferably located so that all dimensions are positive.
3. All dimensions must be in decimals.
4. Angles should be specified by coordinate dimensions rather than degrees.
5. Standard tools, such as drills, reamers, or taps should be specified if known.
6. All tolerances should be determined by the design requirements of the part, and not by the capability of the manufacturing machine.

Checklist for Dimensioning

The following lists many of the situations where a new draftsman is likely to make a mistake when dimensioning. The beginner should check his/her detailed drawing against this list before submitting it to his/her supervisor.

1. Dimensions should be attached to the view where the shape is best shown. (This is the contour principle.)
2. Each dimension should be given clearly so that it can be interpreted in only one way.
3. Dimensions should not be duplicated or the same information be given in two different ways, and no dimensions should be given except those needed to produce or inspect the part.
4. Dimensions should be given between points or surfaces that have a functional relation to each other or that control the location of mating parts.
5. Dimensions should be given so that the machinist does not have to calculate, scale, or assume any dimension.
6. Do not expect the worker to assume a feature is centered (such as a hole in a plate), but rather give a location dimension from one side. However, if the hole is to be centered on a symmetrical rough casting, reference the centerline in a note and omit the locating dimension to the centerline.
7. Avoid dimensioning to hidden lines. Create a sectioned view if necessary.
8. Dimensions applying to two adjacent views should be placed between the views unless clearness is promoted by placing the dimension outside one view.
9. A dimension should be attached to only one view.
10. Dimension lines should be spaced uniformly throughout the drawing. They should be at least 1/2 inch from the part outline and at least 3/8 inch apart.
11. Dimension lines should not cross, if avoidable.
12. The longer dimensions should be placed outside shorter dimensions so that dimension lines do not cross extension lines.
13. Dimension lines and extension lines should not cross; however, extension lines may cross each other.

14. A dimension line should never be drawn through a dimensional value. A dimensional value should never be on top of any line of the drawing.

15. No line of the drawing should be used as a dimension line or coincide with a dimension line.

16. A dimension line should never be joined end-to-end (chain fashion) with any parallel line of the drawing.

17. Dimensional values should be approximately centered between the arrowheads, except in a "stack" of dimensions where the dimensional values are to be "staggered."

18. Dimensional values should never be crowded or in any way made difficult to read.

19. Dimensional values should not be lettered over sectioned areas unless necessary, in which case a clear space should be left for the dimensional value.

20. In engineering drawings, omit all inch marks, except where necessary for clearness, such as 1" VALVE.

21. An overall dimension should be present in the three major directions to give the overall size of the part. The overall dimension can be marked as REF if the other dimensions are needed to meet the design intent.

22. Avoid a complete chain of detail dimensions. Instead, omit one dimension or add REF to one detail dimension or the overall dimension.

23. A centerline may be extended and used as an extension line, in which case it is still drawn like a centerline.

24. Centerlines should not extend from view to view.

25. Leaders for holes should be straight, not curved, and point to the circular views of holes wherever possible.

26. Leaders should slope at an angle between 20 and 70 degrees from the horizontal. Vertical and horizontal leader lines are not acceptable.

27. Leaders should extend from the beginning or end of a note; the horizontal 1/4-inch long "shoulder" should extend from the mid-height of the text.

28. Dimensional values for angles should be listed horizontally.

29. Notes should always be lettered horizontally.

30. Notes should be brief and clear, and the wording should use standard symbols, nomenclature, and abbreviations.

31. Finish marks should be placed on the edge views of all finished surfaces, including hidden edges and the contour and circular views of cylindrical surfaces.

32. Finish marks should be omitted on holes or other features where a note specifies the machining operation.

33. Finish marks should be omitted on parts made from rolled stock.

34. If a part is finished all over, omit all finish marks, and use the general note: FINISH ALL OVER, or FAO.

35. A solid cylinder is dimensioned by giving both its diameter and length in the rectangular view. A diagonal diameter in the circular view may be used in cases where clearness is gained.

36. Holes to be bored, drilled, reamed, etc., are size-dimensioned by notes in which the leaders point toward the circular view of the holes. Shop processes may be omitted from these notes.

37. Drill sizes are preferably expressed in decimals, especially for drills designated by number or letter size.

38. In general, a circle is dimensioned by its diameter, and an arc by its radius. Typically, a circular arc goes through an angle of less than 180 degrees; a circle goes through an angle of 180 degrees or more.
39. A diameter dimensional value should be preceded by the symbol Ø. The letter R should precede the radius dimensional value.
40. Cylinders should be located by their centerlines.
41. Cylinders should be located in the circular views, if possible.
42. Cylinders should be located by coordinate dimensions in preference to angular dimensions when accuracy is important.
43. When there are several rough non-critical features obviously the same size, such as fillets, rounds, ribs, etc., give only a typical dimension, or use a general note.
44. Mating dimensions should be given on the drawings of mating parts.
45. Decimal dimensions should be used when greater accuracy than 1/64 of an inch is required on a machined dimension. Normally 2-place decimals are preferred over fractional dimensions.
46. Avoid cumulative tolerances, especially in limit dimensioning. This can be done by not "chaining" dimensions, but rather specifying each one from a reference datum.

Create a Custom Drawing Template

This book is written assuming the ANSI drafting standard and Third Angle Projection for all drawings, and most of the parts are designed in the IPS (inch-pound-second) system. For ease of printing out your drawings on a standard laser printer only A-size (8.5" x 11") drawings will be made in landscape mode.

This next procedure assumes that you are similar with SOLIDWORKS.

1. Start SOLIDWORKS, if not already in it, then pick the **Advanced...** button.
2. Select the Drawing icon from the Template tab. Pick **OK**.
3. If the box for **Only show standard formats** is checked, uncheck it.
4. Select **A (ANSI) Landscape** from the list of sheet formats. Pick **OK**.
5. Pick the red X in the Model View window because we do not want to attach a part to this drawing template. Sheet 1 is the default first Sheet name. Draw1 is the default drawing name.

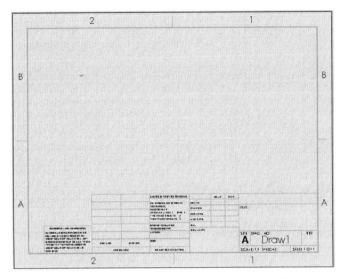

6. In the Feature Manager Design Tree, right-click on Sheet1, then pick **Edit Sheet Format** from the pop-up menu.

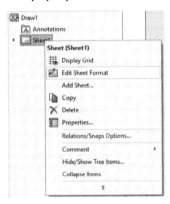

7. Type <Ctrl>-A followed by <Delete> to erase the title block and all its notes. Only the outer border should remain.

8. Under Tools, pick Options. ⚙ Options...

9. Under the System Options tab, under Drawings, select the Display style. Set/verify that the option **Hidden lines visible** is selected.

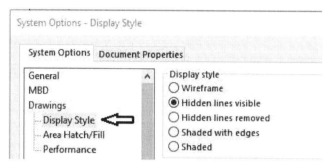

10. Pick the Document Properties tab. Set/verify that ANSI is selected as the Overall Drafting Standard.

11. Pick DimXpert from the list, then select the left-side options. Chamfers are dimensioned as distance and angle and slots are dimensioned center to center as shown below. This will change ANSI to ANSI-MODIFIED.

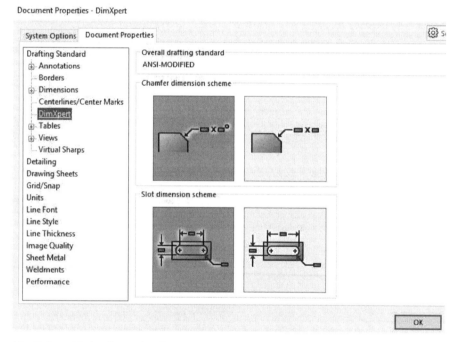

12. Select Units from the list on the left side of the window.

13. Verify that IPS is selected.

14. Set the length to .12, the Dual Dimensions length to .1, and the Angle to .1 by selecting the appropriate option in the pulldown menu. Pick **OK**.

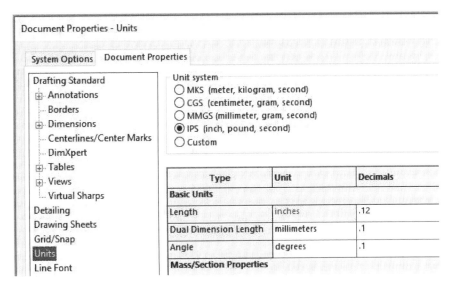

15. In the left-hand column, right-click on Sheet1 to bring up a pop-up menu, then select **Properties...** Verify that the Third Angle is selected and the Scale is 1:1. Select **A (ANSI) Landscape** in the Sheet Format Size list, then pick the **Apply Changes** button.

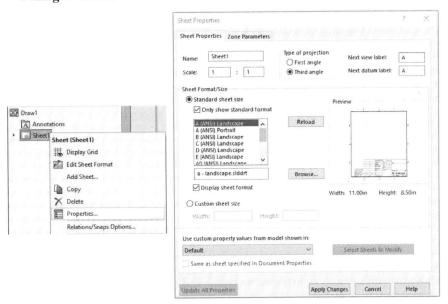

We will use the Document Properties in this template and create a new Sheet Format.

16. Be sure **Enable CommandManager** is selected. If not, right-click at the top of the screen and select it from the pop-up menu.

17. With the Sketch tab selected, draw a rectangle starting in the lower left corner and reaching across to the right side. Dimension the rectangle to be 10 x 1. Make sure the rectangle aligns with the thicker blue border lines. Then use the Line tool to draw horizontal and vertical lines as shown below.

18. Use the Smart Dimension tool to position the lines as shown.

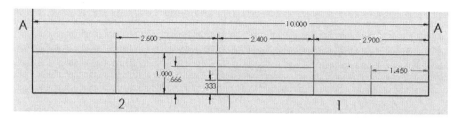

19. Once all the lines are correctly located, delete all of the dimensions.

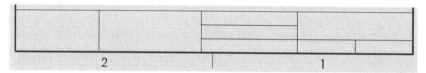

20. 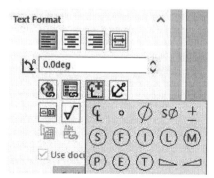 Select the Annotation tab, then pick the Note tool. Enter your company name or your school name in the leftmost area. Select a font and font size. Using Century Gothic size 12 font, I entered:

Ohio Northern University
Mechanical Engr. Dept.
Ada, OH 45810

Pick the red X to close the Formatting window, then center the text in the allocated area. Press the <Esc> key to terminate the note tool.

21. Use the Note tool to enter the following information about units, tolerances, and the drawing standard in all capital letters using Century Gothic size 8 font, enter:

UNLESS OTHERWISE SPECIFIED, DIMENSIONS
ARE IN INCHES, TOLERANCE ARE:
.X±.1   .XX ±.05   .XXX±.001   ANGLES ±.1°

DIMENSIONS & TOLERANCES IN ACCORDANCE
WITH ASME Y14 2009

Use the **Add Symbol** tool to add the plus/minus and the degree symbols to the text. Press the <Esc> key to terminate the note tool.

22. ![A] Use the Note tool to enter Matl: in the top box, Drwn By: in the middle box, and Date: in the bottom box using Century Gothic size 10 font. Press the <Esc> key to terminate the note tool.

23. ![A] Use the Note tool to add Scale: and Dwg No. to the lower right corner areas using Century Gothic size 8 font. Press the <Esc> key to terminate the note tool.

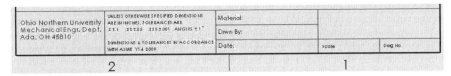

24. Right-click in the graphics area to bring up a pop-up menu. Select Edit Sheet from this menu.
25. Press the "**F**" key to fit the drawing template in the SOLIDWORKS window.
26. Select **Save As…** from the File pull-down menu.
27. In the Save as type: pick **Drawing Template (*.drwdot)**.

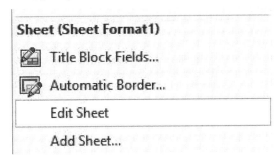

28. Navigate to the **SolidWorks Template** folder that you created in Chapter 1.
29. Enter **A-Size Template** for the filename.
30. Pick the option, **Add a description** beside Description: then type "**A-size ANSI Inch Template.**"
31. Pick the **Save** button.

32. Right-click in the graphics area and pick **Edit Sheet Format** from the pop-up menu.

33. ![A] Use the Note tool to pick the area to the right of Date: in the title block.

34. ![icon] In the Note window, pick the Link to Property icon. When the Link to Property window appears, be sure the **Current document** is selected, then pick **SW-Short Date(Short Date)** from the pulldown menu. Pick **OK**.

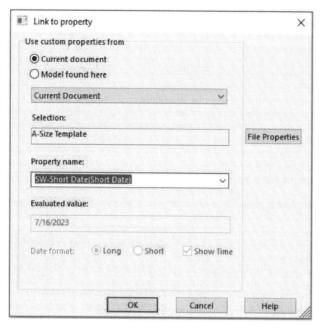

35. Set the Century Gothic font size to 12, then pick the red X in the upper right corner to close the Formatting window. Press the <Esc> key to terminate the note tool.

36. 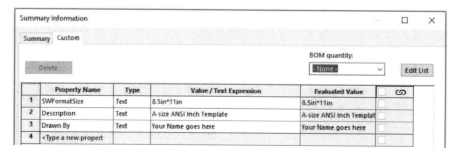 Use the Note tool to pick the area to the right of Drwn By: in the title block.

37. In the Note window, pick the Link to Property icon. When the Link to Property window appears, be sure the **Current document** is selected. Pick the **File Properties** button on the right side of the window.

38. Add **Drawn By** as the Property Name, type as Text, then enter **Your Name** in the Value/Text Expression area. Pick **OK** to close this window.

39. Select Drawn By for the Property Name, then pick **OK**.

40. Set the Century Gothic font size to 10 or less if your name is long, then pick the red X in the upper right corner of the Formatting window. Press the <Esc> key to terminate the note tool.

Summary Information   — □ ×

Summary | Custom

BOM quantity: -None-   Edit List

Delete

	Property Name	Type	Value / Text Expression	Evaluated Value	∞
1	SWFormatSize	Text	8.5in*11in	8.5in*11in	
2	Description	Text	A-size ANSI Inch Template	A-size ANSI Inch Templat	
3	Drawn By	Text	Your Name goes here	Your Name goes here	
4	<Type a new propert				

41. Use the Note tool to pick the area to the right of Material: in the title block.

42. In the Note window, pick the Link to Property icon. When the Link to Property window appears, be sure the **Model found here** is selected. Pick the File Properties button on the right side of the window.

43. Add **Material** as the Property Name, type as Text, then enter **unknown** in the Value/Text Expression area. Pick **OK** to close this window.

44. Select Material for the Property Name, then pick **OK**.

45. Set the Century Gothic font size to 10, then pick the red X in the upper right corner of the Formatting window. Press the <Esc> key to terminate the note tool.

46. Use the Note tool to pick the large area on the right side of the title block.

47. In the Note window, pick the Link to Property icon. When the Link to Property window appears, be sure the **Model found here** is selected.

48. Select **SW-File Name(File Name)** for the Property Name, then pick **OK**.

49. Set the Century Gothic font size to 14, then pick the red X in the upper right corner of the Formatting window. Press the <Esc> key to terminate the note tool.

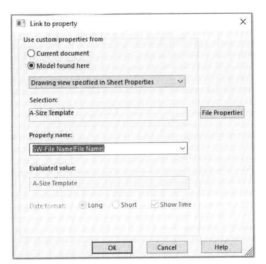

50. Use the Note tool to pick the area to the right of Scale: in the title block.

51. In the Note window, pick the Link to Property icon. When the Link to Property window appears, be sure the **Current document** is selected.

52. Select **SW-Sheet Scale(Sheet Scale)** for the Property Name, then pick **OK**.

53. Set the Century Gothic font size to 12, then pick the red X in the upper right corner of the Formatting window. Press the <Esc> key to terminate the note tool.

54. Use the Note tool to pick the area to the right of Dwg No. in the title block.

55. ▣ In the Note window, pick the Link to Property icon. When the Link to Property window appears, be sure the **Model found here** is selected. Pick the File Properties button on the right side of the window.

	Property Name	Type	Value / Text Expression	Evaluated Value	↺
1	SWFormatSize	Text	8.5in*11in	8.5in*11in	☐
2	Description	Text	A-size ANSI Inch Template	A-size ANSI Inch Templat	☐
3	Drawn By	Text	Your Name Here	Your Name Here	☐
4	Material	Text	unknown	unknown	☐
5	DrwgNo	Text	000000	000000	☐
6	<Type a new propert				☐

56. Add **DrwgNo** as the Property Name, type as Text, then enter **000000** in the Value/Text Expression area. Pick **OK** to close this window.
57. Select **DrwgNo** for the Property Name, then pick **OK**.
58. Set the Century Gothic font size to 12, then pick the red X in the upper right corner of the Formatting window. Press the <Esc> key to terminate the note tool.

Your title block area should appear similar to what is shown below. Don't be concerned with the variable for the drawing number being outside the border. The actual drawing number will be much shorter.

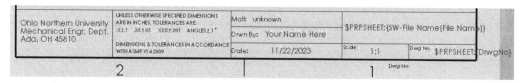

59. Under File, pick **Save Sheet Format...**

60. Navigate to the **SolidWorks Template** folder that you created in Chapter 1.
61. Enter **"A-Size Format.slddrt"** for the filename. Pick the **Save** button.
62. If it states that it already exists. Do you want to replace it? Pick **Yes**.
63. Right-click in the graphics area, then pick **Edit Sheet** from the pop-up menu.
64. Under File, pick **Save** to save any changes to the drawing template.
65. Look in your SolidWorks Template folder on the desktop and verify that the file **"A-Size Template.DRWDOT"** has been updated in this directory.
66. **File> Close**
67. Exit from SOLIDWORKS if stopping at this time.

Creating a Detailed Engineering Drawing

This section describes the procedure necessary to create a detailed drawing for a part using the Drawing Template created in the previous section.

The custom drawing template created assumes that two variables used in the title block are defined by the part. They are **Material** (the material the part is made from) and **DrwgNo** (the drawing number for the detailed print.) If these two variables are not found in the part file, then these two variables will be blank in the title block.

In Chapter 2, problem 2.7 asked you to sketch the **"Support Frame"**. In Chapter 3, problem 3.7 asked you to convert the sketch to a 3D part and assign Cast Alloy Steel as the material. Retrieve this part now, then add the two necessary variables to the part.

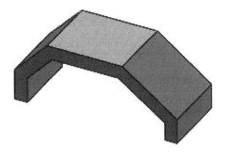

1. ![SW 2024 SOLIDWORKS 2024 icon] Start SOLIDWORKS, if not already in it, then pick the **Advanced...** button.
2. Under File, pick Open
3. Navigate to the SolidWorks Parts folder, select the **"Support Frame"**, then pick the **Open** button.
4. Under File, pick Properties...
5. For the Property Name enter Material. Move to the Value/Text Expression area and enter "SW-Material". The material selected was "Cast Alloy Steel."
6. On the next line enter "DrwgNo", followed by 100307 in the Value/Text Expressions area.
7. Pick **OK** to close the Properties window.

Properties
Summary Custom Configuration Properties Properties Summary

BOM quantity:
- None -

Delete

	Property Name	Type	Value / Text Expression	Evaluated Value
1	Material	Text	Cast Alloy Steel	Cast Alloy Steel
2	DrwgNo	Text	100307	100307
3	\<Type a new propert			

8. Under File, pick Save.
9. Under File, pick Close.

Now we are ready to create a detailed engineering drawing of the Support Frame using the drawing template we created previously.

10. Under File, pick New.
11. Under the SolidWorks Templates tab, pick **A-Size Template**.

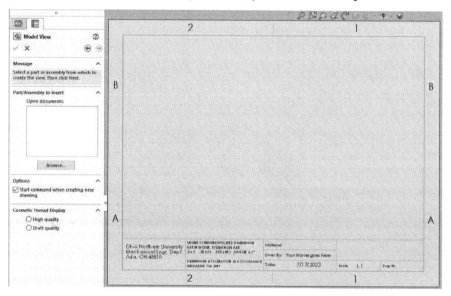

12. Pick the Browse... button, then navigate to your SolidWorks Parts folder.
13. Pick **"Support Frame"** from the list, then pick the Open button.
14. Move the cursor around in the graphics area to place the first view. Use the LMB to place the view.

If the view does not show up, then you saved the A-Size Template in the **Edit Sheet Format** mode. See steps 61 through 64 in the previous section. That is, open A-Size Template, right-click in the graphics area, pick **Edit Sheet** from the popup menu, then save the file, followed by closing the graphics window. Now go back to step 10 above.

15. Move the cursor to the right, then press the LMB to place the right-side view.
16. Move the cursor up to place the 3D view of the part.
17. Press the <Esc> key to exit from creating views.
18. Pick the 3D view and move it to the upper right corner of the drawing area.
19. Under File, pick Save. Be sure to save the drawing in the SolidWorks Parts folder. Name the drawing **"Support Frame Drawing"**.
20. The drawing will be named the same as the part so "Support Frame" should appear in the File name: area on the drawing.
21. Pick the Save button. If it states that **"Support Frame Drawing.SLDDRW"** already exists. Do you want to replace it? Pick the Yes button.
22. Pick the Annotate tab in the upper left corner of the window.

23. 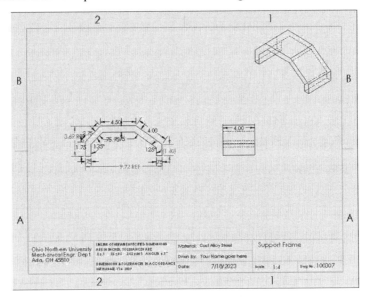 Pick the Model Items icon. In the Source/Destination area, select **Entire model** from the pull-down menu. Pick the green checkmark.

24. Reposition the crowded dimensions. Remove duplicates.

25. Right-click on Sheet Format1, then pick **Properties…**

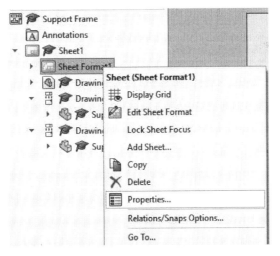

26. Change to scale to 1:3, then pick the **Apply Changes** button.

27. Move the view around so they are similar to the drawing below.

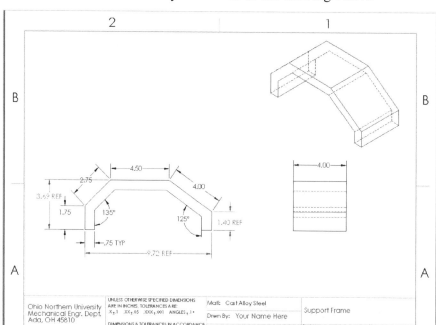

28. 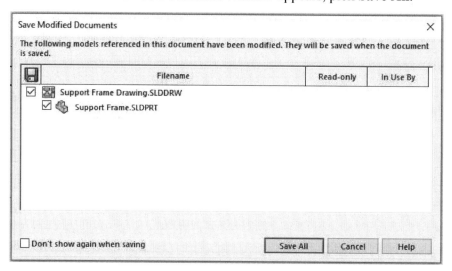 **File> Save** or pick the **Save** icon.
29. When the Save Modified Documents window appears, pick **Save All**.

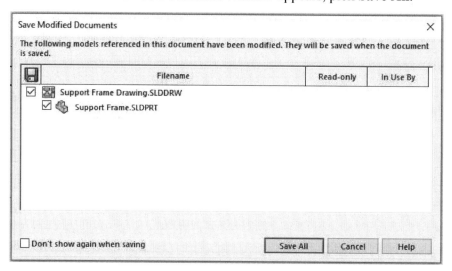

30. If another window comes up with the same filename, pick **Save**.
31. **File> Print...** (if asked to)
32. **File> Close**

Design Intent – Create a slider block with three holes and 2 fillets made from 1060 Alloy. Create the Slider Block based on the sketches below. Use Part_IPS_ANSI under the SolidWorks Template tab as the starting point when creating this part.

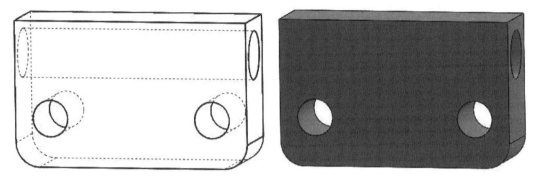

Create a rectangle 3 x 5. Add two ¾ inch holes positioned as shown above. Extrude it midplane 1.5 inches thick. Add a 1-inch hole as shown above that goes through the part. Finally, add two ½ inch fillets to the lower corners. Save the part in the SolidWorks Parts folder as "**Slider Block**", then File> Close.

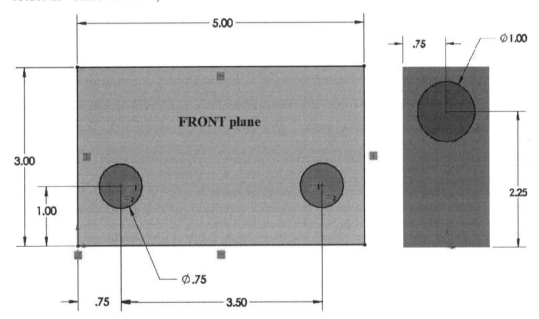

1. **File> New…**
2. Pick the **A-Size Template** under the SolidWorks Template tab. Pick **OK**.
3. Pick the Browse… button, then browse through the SolidWorks Parts folder until you locate the **Slider Block** part, then pick the Open button.
4. Move the cursor to locate the Front view, then press the LMB.
5. Move the cursor to the right to place the RIGHT side view, then press the LMB again. Press <Esc> to terminate the view placement procedure. This time we are not going to create a 3D image of the part in the upper right corner of the drawing.

Note: The Material and DwgNo areas of the title block are blank because we forgot to define these properties for this part.

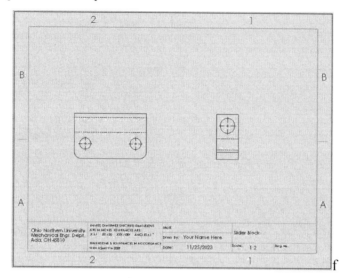

6. Open the Slider Block part, then under File> pick **Properties...**
7. For Property Name enter Material. Move to the Value/Text Expression area and enter "SW-Material". On the second line for Property Name enter DwgNo followed by 200101 in the Value/Text Expression area. Pick **OK**.
8. Save the part.
9. Move back to the drawing.
10. Select the Annotation tab, then pick the Model Items icon.
11. In the Model Item window, set the Source to **Entire model.**
12. In the Dimensions area, make sure the **Marked for Drawing** icon is selected. Pick the green checkmark.

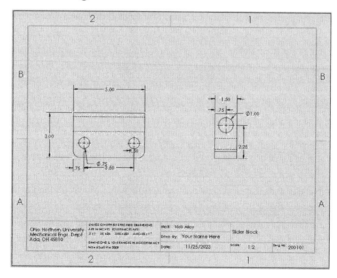

Note that the three holes have centerlines in the view where they are round, but the centerlines are missing in the other view. We need to add two centerlines to the drawing. Be sure to add 2X in front of the hole and fillet dimensions.

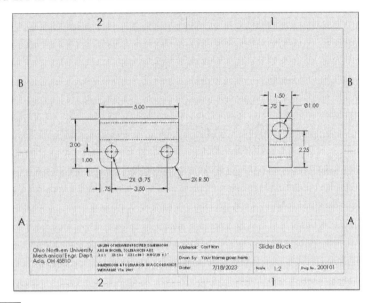

13. 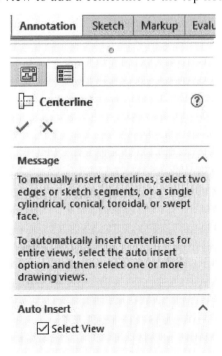 Select the Annotation tab, then pick the Centerline icon.

14. Check the **Select View** box.

15. Pick the Front view to add a centerline to the top hole in the Front view.

16. Pick the Right-side view to add a centerline to the bottom hole.

17. Pick the green checkmark.
18. Now all that is needed is to clean up the dimensions so they are not crowded. Move the dimensions so they are not crowded and look similar to the drawing below. Be sure to add 2X in front of the hole and fillet dimensions and connect the centerlines for the two holes to show they are at the same level.

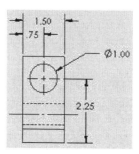

Your final drawing should look similar to the following.

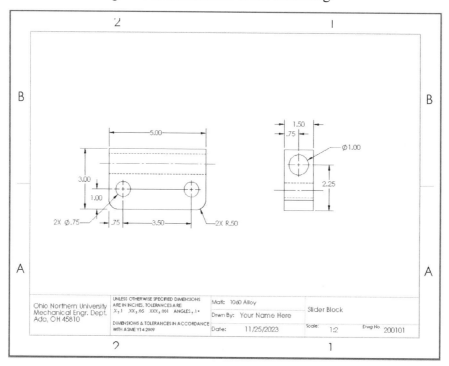

19. **File> Save**
20. When the Save Modified Documents window appears, pick **Save All**.
21. Verify that you are saving the drawing in the SolidWorks Parts folder, then pick **Save**.
22. Enter "**Slider Block Drawing**" as the filename, then pick the **Save** button.
23. **File> Print...** (if asked for)
24. **File> Close**

Creating a Sectioned View

Create the **"Idler Hub"** made from Cast Iron according to the drawing below. Under **File>Properties** define the variable Material as "SW-Material", and the DrwgNo as 210500, then save the part followed by File>Close.

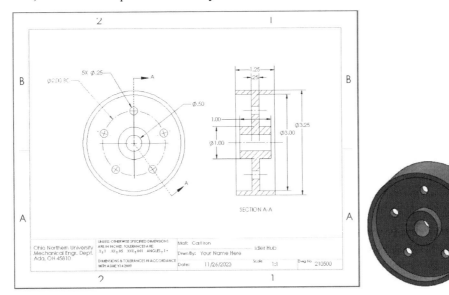

1. **File> New**
2. Pick the **A-Size Template** under the SolidWorks Template tab. Pick **OK**.
3. Browse through the SolidWorks Parts folder until you locate **Idler Hub**, then pick Open.
4. Move the cursor to locate the Front view, then press the LMB followed by the <Esc> key. Your drawing should be similar to the drawing shown below.

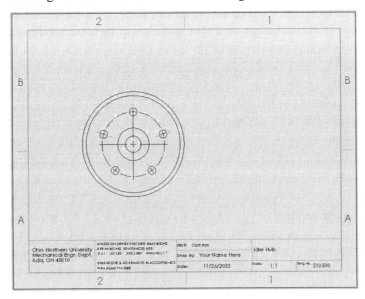

5. 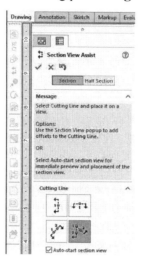 Select the Drawing tab, then pick the **Section View** icon. Check the **Auto-start section view** box. Also, pick the Cutting plane **Aligned** icon.

6. Move the cursor to the center of the hub, then press the LMB (left mouse button).
7. Align the upper position of the cutting plane line with the center of the upper hole, then press the LMB.
8. Align the lower position of the cutting plane line with the center of the lower right-side hole, then press the LMB.
9. Move the cursor to the right to position the sectioned view. Press the LMB to place it.

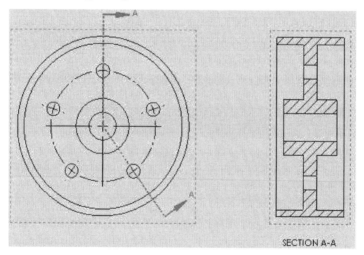

10. 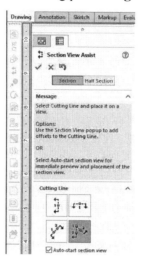 Select the Annotation tab, then pick the Centerline icon.
11. Check the **Select View** box.
12. Pick the Right-side view to add a centerline to the three holes.
13. Pick the green checkmark.

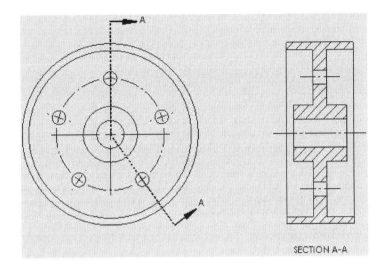

14.  Select the Model items icon. In the Source/Destination window, pick the **Entire model.** Pick the green checkmark to place the dimensions used in creating the idler hub on the drawing.

15. Using the Smart Dimension tool finish adding dimensions to the drawing if necessary. Your drawing should look similar. Use the LMB to select the 2.00-inch diameter bolt circle, then pick the Leader tab and select the diameter icon to change the radius dimension to a diameter dimension, then add " BC" to the dimensional value.

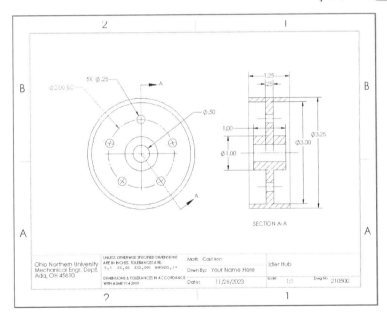

16. Save the drawing as **"Idler Hub Drawing"** and the part if asked.
17. **File> Print…** (if asked to)
18. **File> Close**

Creating an Auxiliary View

The following part has a slanted surface so none of the standard views (Front, Top, and Right) will show the hole in the part as a circle. An auxiliary view perpendicular to the slanted surface is needed.

The **"Angle Brace"** made from plain carbon steel was problem 3.11. It has a slanted surface with a hole in it so an auxiliary view is needed to accurately show the hole's size and shape.

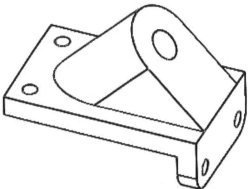

If problem 3.11 was not assigned, create the **"Angle Brace"** now according to the drawing below. Under File> pick **Properties...** For Property Name enter Material. Move to the Value/Text Expression area and enter "SW-Material". On the second line for Property Name enter DrwgNo followed by 111031 in the Value/Text Expression area, then save the part followed by File>Close.

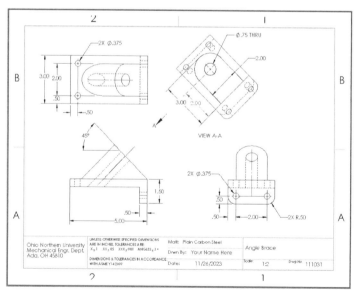

1. **File> New...**
2. Pick the **A-Size Template** under the SolidWorks Template tab. Pick **OK**.
3. Locate **"Angle Brace"** in the SolidWorks Parts folder, then pick Open.

4. Move the cursor to locate the Front view, then press the LMB. Move the cursor to the right, then press the LMB to place the right-side view. Move the cursor above the Front view, then press the LMB to place the Top view followed by the <Esc> key. Your drawing should be similar to the drawing shown below.

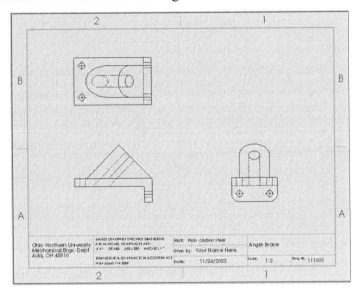

5. 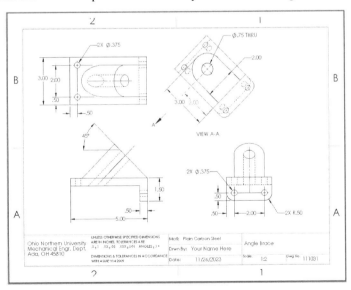 Select the Drawing tab, then pick the Auxiliary View icon.

6. Pick the slanted surface in the Front view, then move the cursor perpendicular to this surface. Press the LMB to place the auxiliary view. Pick the green checkmark.

7. Reposition the views if necessary.
8. Add dimensions to the drawing as needed
9. Save the drawing as **"Angle Brace Drawing"** and the part if asked.
10. **File> Print…** (if asked to)
11. **File> Close**

Dimensioning Problems

6-1   Make a detailed engineering drawing of the **"Rocker"**, problem 3.2. Material = Malleable Cast Iron, and DrwgNo = 200302

6-2   Make a detailed engineering drawing of the **"Latch Plate"**, problem 3.3. Material = 1060 Alloy, and DrwgNo = 200303

6-3   Make a detailed engineering drawing of the **"Cover Plate"**, problem 3.5. Material = Plain Carbon Steel, and DrwgNo = 200305

6-4   Make a detailed engineering drawing of the **"Front Cover Plate"**, problem 3.6. Material = Plain Carbon Steel, and DrwgNo = 200306

6-5   Make a detailed engineering drawing of the **"Breather Plate"**, problem 3.9. Material = Plain Carbon Steel, and DrwgNo = 200309

6-6   Make a detailed engineering drawing of the **"Hanging Bracket"**, problem 3.12. Material = Malleable Cast Iron, and DrwgNo = 200312

6-7   Make a detailed engineering drawing of the **"U-Brace"**, problem 3.13. Material = 1060 Alloy, and DrwgNo = 200313

6-8   Make a detailed engineering drawing of the **"Bearing Plate"**, problem 3.14. Material = Plain Carbon Steel, and DrwgNo = 200314

6-9   Make a detailed engineering drawing of the **"Slotted Cam"**, problem 3.15. Material = Cast Alloy Steel, and DrwgNo = 200315

6-10   Make a detailed engineering drawing of the **"Operating Arm"**, problem 3.16. Material = Plain Carbon Steel, and DrwgNo = 200316

6-11   Make a detailed engineering drawing of the **"Cutoff Holder"**, problem 3.17. Material = Malleable Cast Iron, and DrwgNo = 200317

6-12   Make a detailed engineering drawing of the **"Bearing Holder"**, problem 3.18. Material = Plain Carbon Steel, and DrwgNo = 200318

6-13   Make a detailed engineering drawing of the **"Clutch Lever"**, problem 3.19. Material = Plain Carbon Steel, and DrwgNo = 200319

6-14   Make a detailed engineering drawing of the **"Vibrator Arm"**, problem 3.21. Material = Malleable Cast Iron, and DrwgNo = 200321

6-15   Make a detailed engineering drawing of the **"Operating Lever"**, problem 3.22. Material = Malleable Cast Iron, and DrwgNo = 200322

6-16   Make a detailed engineering drawing of the **"Clutch Cone 15 hp"**, problem 4.1. Material = Plain Carbon Steel, and DrwgNo = 200401

6-17   Make a detailed engineering drawing of the **"Step Pulley"**, problem 4.2. Material = AISI 1020 Cold-rolled Steel, and DrwgNo = 200402

6-18   Make a detailed engineering drawing of the **"Axle"**, problem 4.4. Material = Plain Carbon Steel, and DrwgNo = 200404

6-19   Make a detailed engineering drawing of the **"Offset Shaft"**, problem 4.5. Material = AISI 4340 Annealed Steel, and DrwgNo = 200405

6-20   Make a detailed engineering drawing of the **"V-belt Pulley-6.0"**, problem 4.6. Material = Malleable Cast Iron, and DrwgNo = 200406

6-21   Make a detailed engineering drawing of the **"V-belt Idler"**, problem 4.7. Material = AISI 1010 Hot rolled Bar Steel, and DrwgNo = 200405

6-22   Make a detailed engineering drawing of **"Bearing 232"**, problem 5.1. Material = AISI 4340 annealed Steel, and DrwgNo = 200501

6-23   Make a detailed engineering drawing of the **Female Sleeve"**, problem 5.2. Material = Malleable Cast Iron, and DrwgNo = 200502

6-24   Make a detailed engineering drawing of the **"Fixed Bearing Cup"**, problem 5.3. Material = AISI 4130 annealed Steel, and DrwgNo = 200503

6-25   Make a detailed engineering drawing of the **"End Plate"**, problem 5.5. Material = Plain Carbon Steel, and DrwgNo = 200505

6-26   Make a detailed engineering drawing of the **"Pipe Plug"**, problem 5.8. Material = Malleable Cast Iron, and DrwgNo = 200508

6-27   Create the **"Rod Guide"** made from Plain Carbon Steel, then create a dctailed engineering drawing similar to the two views shown below. Add a general note stating "All rounds and fillets are R.06" and DrwgNo = 200627

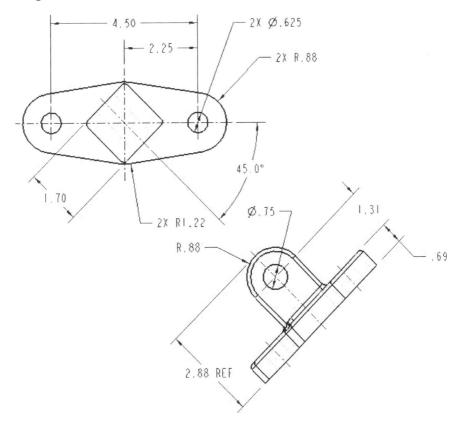

6-28   Create the **"Clamp Fixture"** made from AISI 1020 Steel, then create a detailed engineering drawing similar to the two views shown below with DrwgNo = 200628

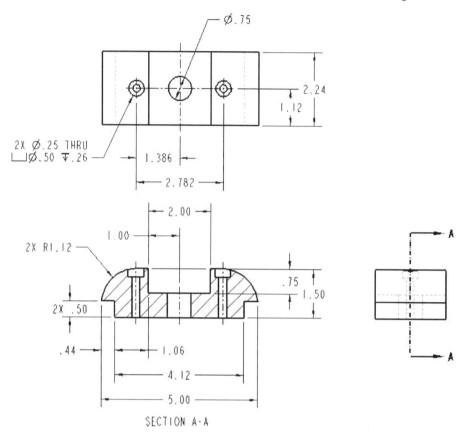

6-29   Create the **"Fixed Bearing Cup"** made from Plain Carbon Steel, then create a detailed engineering drawing with a sectioned view. DrwgNo = 200629

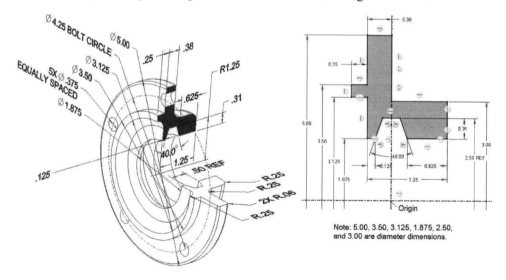

Note: 5.00, 3.50, 3.125, 1.875, 2.50, and 3.00 are diameter dimensions.

6-30 Create the **"Truck Wheel"** made from Malleable Cast Iron, then create a detailed engineering drawing with a sectioned view. DrwgNo = 200630

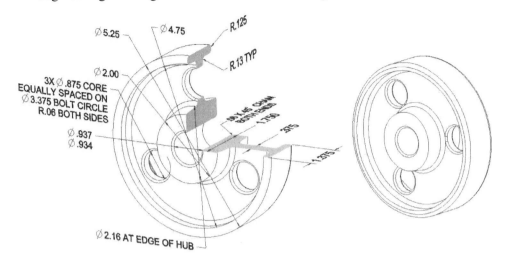

6-31 Create the **"Four-hole Hub"** made from Plain Carbon Steel, then create a detailed engineering drawing similar to the two views shown below with DrwgNo = 200631

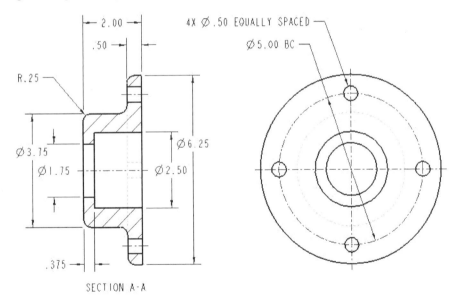

6-32   Design the **"Guide Bracket_302"** made from Cast Alloy Steel according to the figures below. The part is 0.50 inches thick. Is a third view necessary? **Design change:** the four mounting holes need to be .53 inches in diameter. Then make a detailed engineering drawing of the part with drawing number 160111.

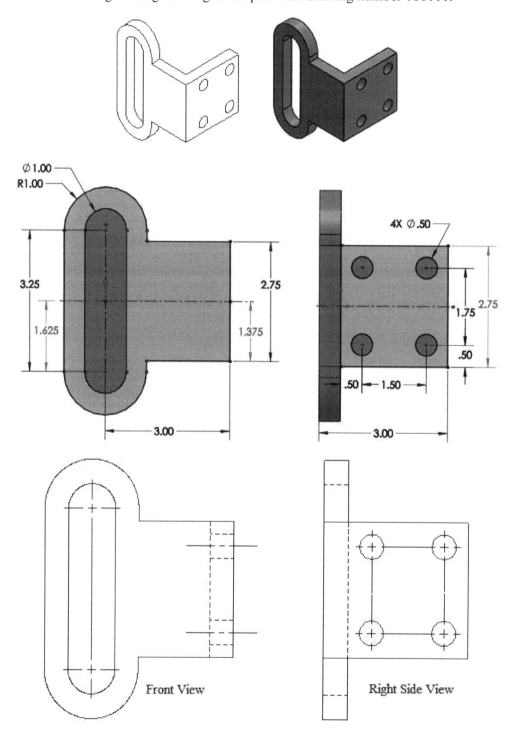

6-33   Design the **"Support Bracket_305"** made from Cast Alloy Steel according to the figures below. The part is 0.50 inches thick. Is a third view necessary? **Design change:** the four mounting holes need to be .53 inches in diameter. Make a detailed engineering drawing of the part with drawing number 160121.

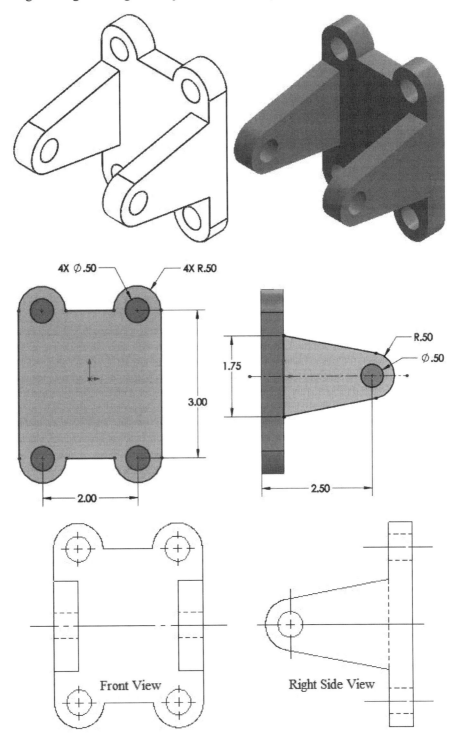

6-34   Design a **"Heat Sink"** similar to the one shown here, but made from a solid aluminum extrusion (6061 Alloy) instead of a group of pieces bolted together. The base is 1.81 inches by 3.00 inches with a thickness of 0.375 inches. Each fin is 0.06 inches thick, 3.00 inches long, and protrudes above the base plate by 1.25 inches. There is a 0.19-inch gap between each of the eight fins. After creating one fin, create the other fins using the linear pattern tool.

Next, make a detailed engineering drawing of the Heat Sink with DrwgNo = 200632.

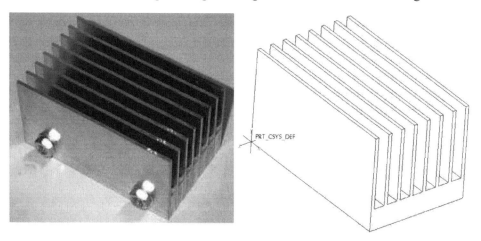

6-35   Design the **"Slanted Block"** made from Plain Carbon Steel according to the figures below. Then make a detailed engineering drawing of the part with drawing number 200633. Use either a sectioned view or an auxiliary view in the drawing.

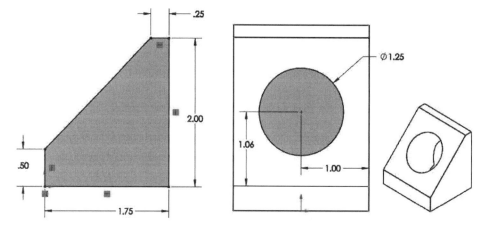

Chapter 7.   Parametric Modeling using Equations

Introduction

In SOLIDWORKS dimensional values are used to size or locate geometric features. The Smart Dimension tool is used to add dimensions to a sketch. Each dimension is assigned a name so that it can be referenced by other dimensions if necessary. The default name is "Dxx" where "xx" is a number that SOLIDWORKS increments automatically each time a new dimension is added. The full name of the dimension is "Dxx@entity" where "entity" is where the dimension is applied. For example,"D10@Sketch2" is the name of the tenth dimension added to Sketch2.

Let's look at a **"Pipe Support"** where all the size dimensions are based on the Pipe Diameter being 2.50 inches.

0.375 = upper edge width = 15% of Pipe Diameter
0.75 = thickness of lower portion = 30% of Pipe Diameter
1.50 = height above lower portion = 60% of Pipe Diameter
7.25 = overall length = 290% of Pipe Diameter
3.00 = depth for extrusion = 120% of Pipe Diameter
0.75 = mounting hole diameters = 30% of Pipe Diameter
1.00 = location of mounting hole from left end = 40% of Pipe Diameter
5.25 = distance to 2nd hole from first hole = 210% of Pipe Diameter =

1. Let's begin by starting SOLIDWORKS. Be sure to toggle into **Advanced** mode. Select the **Part_IPS_ANSI** SOLIDWORKS Template, then pick **OK**.
2. Create a property variable by selecting **File> Properties…**
3. Type **"Pipe Diameter"** as the Property Name. Select **Number** as the Type. Enter **2.5** as the value. Pick **OK**.

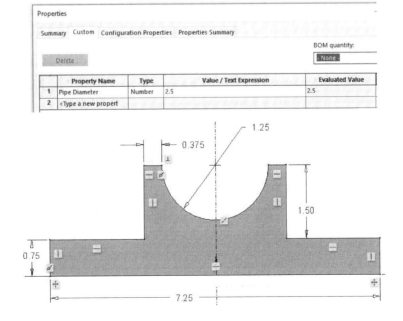

4. Select the Front plane, then pick the sketch icon.
5. Sketch the following symmetrical shape shown on the previous page.
6. Double-click on the overall length dimension of 7.25 and enter the appropriate equation. Start with the equal sign and type: =**"Pipe Diameter"*2.9**

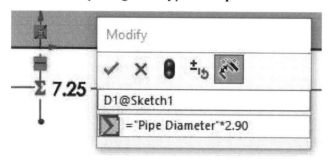

7. Do the same for all of the other dimensions as shown based on the design intent. The summation sign (Σ) indicates that the dimensional value is controlled by an equation. Instead of typing "pipe Diameter" every time, you can pick **Pipe Diameter** out of the File Properties pop-up menu, then add the **"*x.xx"** value without the quote marks.

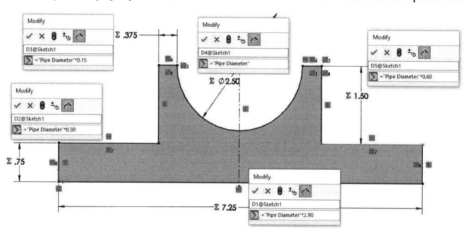

8. Exit Sketch.

9. Pick the Extruded Boss/Base icon. For Direction 1, pick **Mid Plane** from the pulldown menu. Enter the equation = **"Pipe Diameter"*1.2** for the depth of the extrusion., then pick the green checkmark.

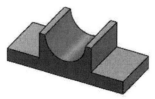

10. [icon] Pick the Sketch icon, then pick either top surface of the extended part.
11. Use the Circle tool to draw two circles, one on each side of the center.
12. Use the Smart Dimension tool to size and locate the left hole.
13. Add a relation so that the second hole is the same size and located on the same horizontal line.
14. Use the Smart Dimension tool to define the distance between the two holes.
15. Double-click on each dimension and enter the appropriate equation based on design intent. The summation sign (Σ) indicates that the dimensional value is controlled by an equation.

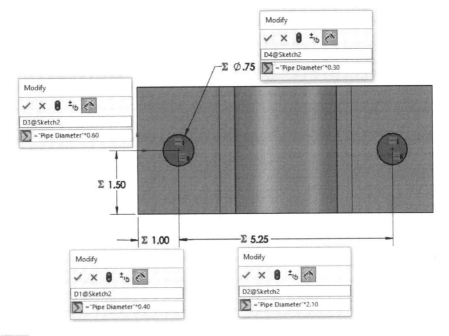

16. [icon] Exit Sketch.

17. [icon] Select the Extruded Cut icon. Select **Through All**, then pick the green checkmark.

18. Reposition the Pipe Support.

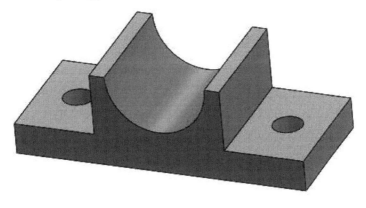

19. 🖫 Save the part in your SolidWorks Parts folder. Name it **"Pipe Support"**.

Now let's check to see if this pipe support meets our design intent by changing the pipe diameter to 2.00 inches.

20. File > Properties…
21. Change the property variable, Pipe Diameter, to 2.00 inches, then pick OK.

22. 🔘 Pick the **Rebuild** icon at the top of the window. The pipe support should become smaller on the screen but remain exactly the same shape. If it changes shape in any way, then go back and verify the equations used with your dimensions.
23. File > Properties…
24. Change the property variable, Pipe Diameter, to 1.00 inches, then pick OK.

25. 🔘 Pick the **Rebuild** icon to update the part again.
26. Edit Sketch1. Your sketch should have the following dimensional values.

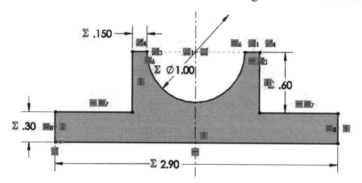

27. Exit Sketch.
28. File Properties…
29. Change the property variable, Pipe Diameter, back to 2.50 inches. Pick OK.

30. 🔘 Pick the **Rebuild** icon to update the part again.

Note that there is a new item in the Feature Manager Design Tree called Equations.

31. Right-click on Equations to bring up a pop-up menu. Pick **Manage Equations…** from the list.

The Equations, Global Variables, and Dimensions window appears. It shows all of the dimensional values controlled by equations and their current value. Note that the property value "Pipe Diameter" and its current value of 2.50 are not shown in this list.

Instead of creating Pipe Diameter as a property variable, it could have been defined as a global variable in which case it would be shown in the Global Variables section along with its current value.

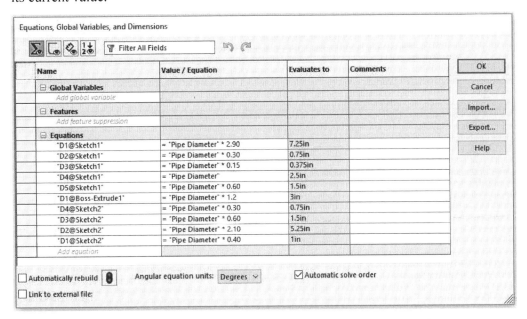

Name	Value / Equation	Evaluates to	Comments
⊟ **Global Variables**			
Add global variable			
⊟ **Features**			
Add feature suppression			
⊟ **Equations**			
"D1@Sketch1"	= "Pipe Diameter" * 2.90	7.25in	
"D2@Sketch1"	= "Pipe Diameter" * 0.30	0.75in	
"D3@Sketch1"	= "Pipe Diameter" * 0.15	0.375in	
"D4@Sketch1"	= "Pipe Diameter"	2.5in	
"D5@Sketch1"	= "Pipe Diameter" * 0.60	1.5in	
"D1@Boss-Extrude1"	= "Pipe Diameter" * 1.2	3in	
"D4@Sketch2"	= "Pipe Diameter" * 0.30	0.75in	
"D3@Sketch2"	= "Pipe Diameter" * 0.60	1.5in	
"D2@Sketch2"	= "Pipe Diameter" * 2.10	5.25in	
"D1@Sketch2"	= "Pipe Diameter" * 0.40	1in	
Add equation			

32. Pick **OK** to close this window.
33. In the Feature Manager Design Tree select **Material <not specified>**, then pick **Plain Carbon Steel** from the list.
34. Save the part.
35. **File> Close**

Note: An important part of parametric modeling is controlling dimensional values with equations. You can also rename the dimensions using names that pertain to their function instead of using the default name "Dxx".

You can create a global variable that controls all or part of the dimensions of the part. This allows you to meet the design intent for any specified size.

---

1. If you exited from SOLIDWORKS, restart it now. Be sure to toggle into **Advanced** mode. Select the **Part_IPS_ANSI** SOLIDWORKS Template, then pick **OK**.

2. Select the Front plane, then pick the sketch icon.
3. Sketch the following shape, then add the appropriate dimensions using the Smart Dimension tool. The original rectangle is centered at the origin.

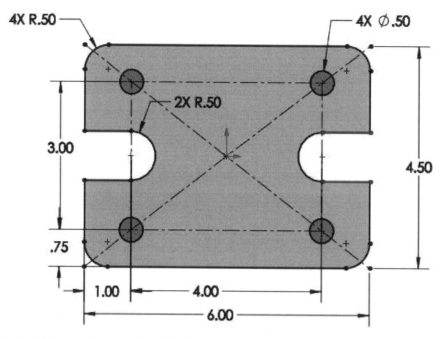

4. Right-click on the arrow beside the eyeball in the View Heads Up toolbar. Pick the D1 icon so the dimensional variable names will show up on the sketch. Then one at a time change the default dimensional name "Dxx" to the names shown below so that all dimensional names represent the feature they are associated with.

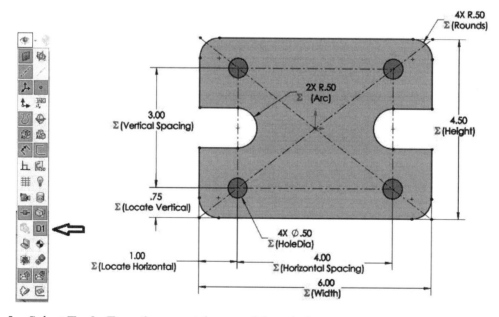

5. Select **Tool> Equations…** at the top of the window.
6. Under the Global Variables area enter: "Size" as the Global Variable name and "6" as its value.

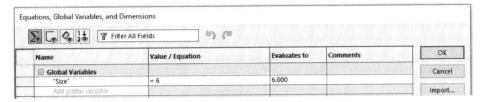

7. Rearrange the window to the sketch containing the dimensional names and the Equations, Global Variables, and Dimensions window are both visible.
8. Pick the first area under the Equations section.
9. Pick the "Width" dimension on the sketch. In the Value/Equation area beside it, enter "Size". Pick the green checkmark. (If the checkmark doesn't show up, then there is an error in your equation.)
10. Move to the next line, then pick the "Height" dimension from the sketch followed by "size"*0.75. Pick the green checkmark.
11. Then move to the next line. Repeat this process until all the dimensions have the appropriate equation associated with them.

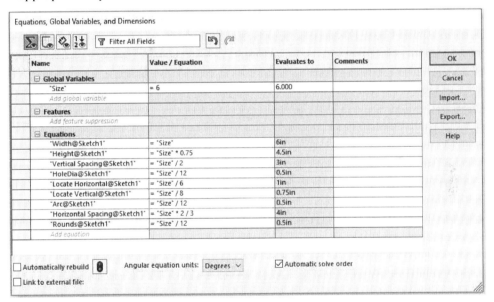

12. Pick **OK** to close the window.
13. All of the dimensional names should have a summation sign (Σ) in front of them. If not, right-click on Equations in the Feature Manager Design Tree to open the window again.
14. Pick the Rebuild icon at the top of the window. The shape of the sketch should not change if all equations were entered correctly.
15. Exit Sketch.

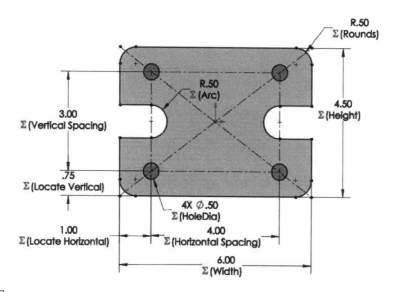

16.  With Sketch1 highlighted, pick the Extruded Boss/Base icon.
17. For depth of the extrusion enter "Size"/30, then pick the green checkmark.

18. In the Feature Manager Design Tree select **Material <not specified>**, then pick **Plain Carbon Steel** from the list.
19. Save the part. Name it **"Cover Plate"**.
20. Right-click on Equations in the Feature Manager Design Tree. Pick **Manage Equations...** from the pop-up menu to open the Equations, Global Variables, and Dimensions window again. Note that a new variable has been added to the list relating the depth of the extrusion to the Global variable named "Size".
21. Change the global variable "Size" to 12.
22. Pick OK to close this window.
23. Edit Sketch1
24. The sketch should look similar to the following.

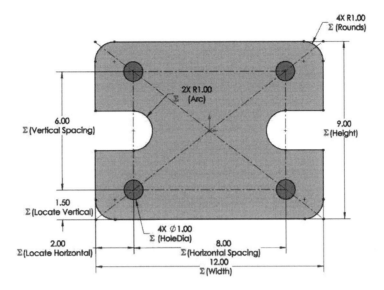

25. Exit sketch.

26. Right-click on Equations in the Feature Manager Design Tree. Pick **Manage Equations…** from the pop-up menu to open the Equations, Global Variables, and Dimensions window again. Note that a new variable has been added to the list relating the depth of the extrusion to the Global variable named "Size".

27. Change the global variable "Size" back to 6.

28. Pick OK to close this window.

29. Save the part.

30. **File> Close**

Note that if the design intent for this part or any part defined in this way changes, changing the defining equations will alter the design according to the new design intent.

Although this example changed all the dimensional variable names, this is not necessary. The procedure works the same using the default variable name "Dxx" except it may not be clear what equation changes in the design.

Design Intent, Using Equations, and Patterns

Design Intent: For roller chain sizes #40 through #160, create a corresponding sprocket with a specified number of teeth. The sprocket will have a specified hole diameter at its center. The pitch in inches is equal to the Chain size divided by 80. The thickness of the sprocket is 0.57 times the pitch. The pitch diameter is the pitch divided by the sine of the quantity 180 degrees divided by the number of teeth on the sprocket. A hub diameter will also be specified. If the hub diameter is equal to zero, then two ¼-inch mounting holes will be located on a bolt circle equal to twice the hole diameter minus ⅛ inch.

1. If you exited from SOLIDWORKS, restart it now. Be sure to toggle into **Advanced** mode. Select the **Part_IPS_ANSI** SOLIDWORKS Template, then pick **OK**.
2. Select **Tool> Equations…** at the top of the window.
3. Under the Global Variables area, enter: "Teeth" as the Global Variable name and **18** as its value.
4. On the next line, enter "ChainNo" as the name and **60** as its value.
5. On the next line, enter "ShaftDia" as the name and **0.875** as its value.
6. On the next line, enter "HubDia" as the name and **2.00** as its value.
7. On the next line, enter "Pitch" followed by **"ChainNo"/80**
8. On the next line, enter "PitchDia" followed by **"Pitch"/sin(180 / "Teeth")**
9. On the next line, enter "DiaRoller" followed by **0.628* "Pitch" + 0.012**

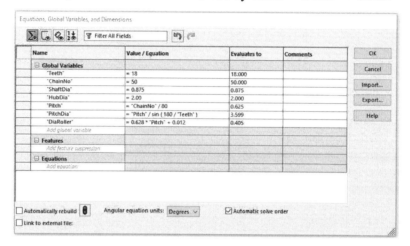

10. Pick **OK** to close the window.
11. 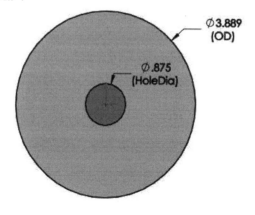 Select the Front plane, then pick the sketch icon.
12. Sketch the following shape, then add the appropriate dimensions using the Smart Dimension tool.

13. **D1** Right-click on the arrow beside the eyeball in the View Heads Up toolbar. Pick the **D1** icon so the dimensional variable names will show up on the sketch. Rename the larger diameter dimension's name as **"OD"** and the smaller diameter dimension's name as **"HoleDia"**.

14. Double-click on the outside diameter to bring up the Modify window. Enter after the equal sign: **"PitchDia" + 0.625*"Pitch" – 0.25*"DiaRoller"**, then pick the green checkmark.

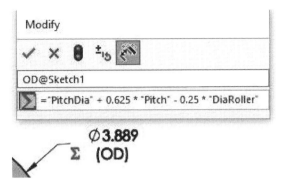

15. Right-click on Equations, then pick **Manage Equations…**
16. Select the 2nd line under the Equations area, then pick the **HoleDia** dimension in the graphics area. In the Value/Equation area, in the Global Variables pulldown menu, pick **ShaftDia**. Pick the green checkmark, then pick **OK** to close the window.

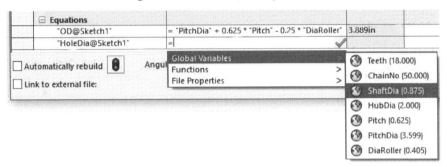

The sketch should have a summation sign (Σ) in front of both dimensions along with the values of 3.889 and .875 as shown below.

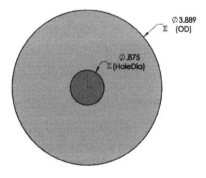

17. ⌐ Exit sketch.

18. 🗔 With **Sketch1** highlighted, pick the Extruded Boss/Base icon.

19. For the depth of the Blind extrusion enter after the equal sign: **"Pitch"*0.57**, then pick the green checkmark after the equation, then pick the green checkmark to accept the extruded depth.

20. ⊟ Save the Part. Name it **"Sprocket Design"**.

21. ⊏ Pick the front surface of the sprocket blank, then pick the Sketch icon.

22. ⊙ Draw a construction circle centered at the origin.

23. Use the Smart Dimension tool to define its diameter.

24. Right-click on Equations, then pick **Manage Equations…**

25. Select the next blank line under the Equations area, then pick the **D1** dimension in the graphics area. In the Value/Equation area, in the Global Variables pulldown menu, pick **PitchDia**. Pick the green checkmark, then pick **OK** to close the window.

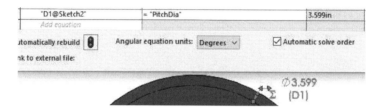

26. ⊿ Draw a vertical centerline through the origin.

27. ⊙ Draw a circle centered where the vertical centerline and the pitch diameter circle meet.

28. Draw two construction lines each 62.85 degrees away from the vertical centerline.

29. ✄ Trim the portion of the circle above these two construction lines.

30. Dimension the radius of the remaining arc to be =**0.272*"Pitch"+0.036**

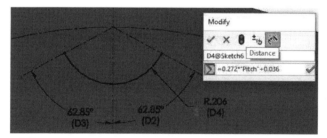

31. ⌒ Use the 3-point arc to draw two arcs that are tangent to the R.206 arc and equal in size. Then dimension the arc to be =**2.07*"Pitch"+0.255**

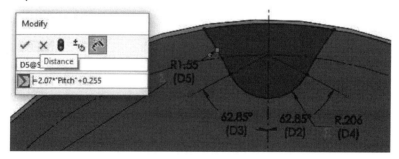

32. Use the 3-point arc to enclose the area. The radius of this arc should be set to =**"OD@sketch1"/2** and with 3-decimal places is equal to 1.944 inches.

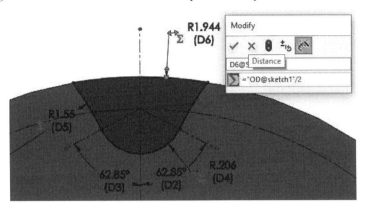

33. Exit sketch.

34. With **Sketch2** highlighted, pick the Extruded Cut icon.

35. For Direction 1 pick **Through All**, then pick the green checkmark to accept the extruded cut.

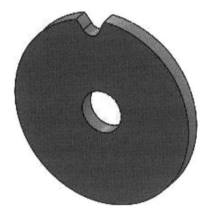

36. With **Cut-Extrude1** highlighted, pick the Circular pattern icon.

37. For Direction 1, pick the small hole circle.

38. With equal spacing selected, enter =**"Teeth"** for the number of occurrences in 360 degrees.

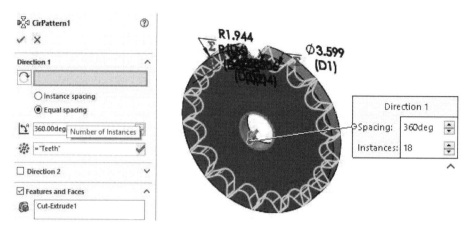

39. Pick the green checkmark to accept the circular pattern.

40.  Save the part.

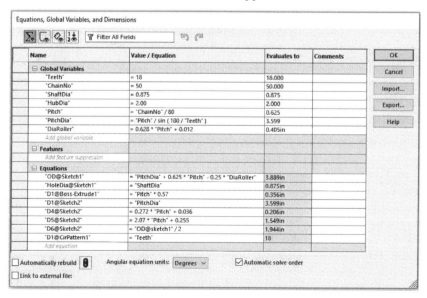

41. Right-click on Equations, then pick **Manage Equations...** The Equations, Global Variables, and Dimensions window should appear similar to what is shown below.

Name	Value / Equation	Evaluates to	Comments
Global Variables			
"Teeth"	= 18	18.000	
"ChainNo"	= 50	50.000	
"ShaftDia"	= 0.875	0.875	
"HubDia"	= 2.00	2.000	
"Pitch"	= "ChainNo" / 80	0.625	
"PitchDia"	= "Pitch" / sin (180 / "Teeth")	3.599	
"DiaRoller"	= 0.628 * "Pitch" + 0.012	0.40Sin	
Add global variable			
Features			
Add feature suppression			
Equations			
"OD@Sketch1"	= "PitchDia" + 0.625 * "Pitch" - 0.25 * "DiaRoller"	3.889in	
"HoleDia@Sketch1"	= "ShaftDia"	0.875in	
"D1@Boss-Extrude1"	= "Pitch" * 0.57	0.356in	
"D1@Sketch2"	= "PitchDia"	3.599in	
"D4@Sketch2"	= 0.272 * "Pitch" + 0.036	0.206in	
"D5@Sketch2"	= 2.07 * "Pitch" + 0.255	1.549in	
"D6@Sketch2"	= "OD@sketch1" / 2	1.944in	
"D1@CirPattern1"	= "Teeth"	18	
Add equation			

We are not done yet. The design intent allows for a hub or two mounting holes to be included as part of the design. If the hub diameter is zero, then no hub is present in which case there will be two ¼ inch mounting holes added to the sprocket. Let's create the hub first.

42. Select the front surface of the sprocket, then pick the sketch icon.

43. Draw a circle centered at the origin about twice the diameter of the center hole. Use the Smart Dimension tool to dimension its diameter, then set its diameter equal to the global variable, **"HubDia"**. Pick the green checkmark twice.

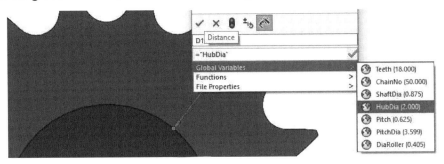

44. Draw a second circle centered at the origin equal to the diameter of the center hole. Use the Smart Dimension tool to dimension its diameter, then set its diameter equal to the global variable, **"ShaftDia"**. Pick the green checkmark twice.

45. Exit sketch.

46. With **Sketch3** highlighted, pick the Extruded Boss/Base icon.

47. For the depth of the Blind extrusion enter 1.00 inches, then pick the green checkmark to accept the extruded depth.

48. Use the Hole tool to add a threaded hole to the hub. Pick the outer surface of the hub near the top.

49. Straight Tap Select ¼-20 for the thread size. Select **Up to Next** for the End Condition for the hole and the thread.

50. Pick the Position tab in the Hole Specification window. Pick a point on the top of the hub. Using the Smart Dimension tool, pick the center of the threaded hole. Enter x = 0, y = 1, and z = 1, then pick the green checkmark twice.

51. Save the part.

52. Highlight **Boss-Extrude2** and **¼-20 Tapped Hole1** in the Feature Manager Design Tree, then right-click and select **Configure Feature** from the pop-up menu. Check the Suppress box for both features, then pick **OK**. The hub and threads are suppressed.

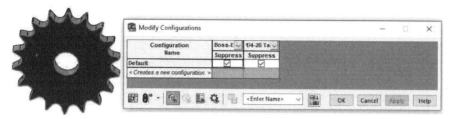

53.  Select the front surface of the sprocket, then pick the Sketch icon.

54. Use the circle tool to draw a construction circle, then set its diameter to the **="ShaftDia"*2 – 1/8**.

55. Draw a vertical centerline through the origin.

56. Draw two small circles where the larger circle intersects with the vertical centerline. Add a relation so both small circles are the same size. Then set the diameter of the two circles equal to 0.26 inches.

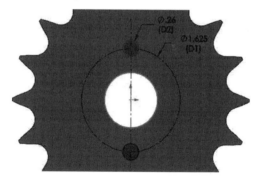

57. Exit sketch.

58. With **Sketch5** highlighted, pick the Extruded Cut icon.

59. For Direction 1 pick **Through All**, then pick the green checkmark to accept the extruded cut.

60. Save the part.

61. Right-click on Equations in the Feature Manager Design Tree, then pick **Manage Equations…**

62. Add the three lines to the Suppress area as shown below, then pick **OK**.

⊟ Features		
"Cut-Extrude2"	= IIF ("HubDia" > "ShaftDia" , "suppresse	Suppressed
"Boss-Extrude2"	= IIF ("HubDia" > "ShaftDia" , "unsuppres	Unsuppressed
"1/4-20 Tapped Hole1"	= IIF ("HubDia" > "ShaftDia" , "unsuppres	Unsuppressed

The sprocket should appear with the hub previously defined and the two mounting holes should not be visible because the hub diameter is larger than the shaft diameter.

63. Right-click on Equations in the Feature Manager Design Tree, then pick **Manage Equations…**

64. Change the global variable, **HubDia** to 0.5, then pick **OK.**

The hub disappears and the two mounting holes reappear.

When we compare our design with an actual sprocket we note that there is a 0.03 x 45° chamfer on each tooth. We do not want to delete all our work to add the chamfer before we cut the teeth. Luckily, SOLIDWORKS allows you to back the design up without losing any of your work.

65. Select the blue line at the end of the **Feature Manager Design Tree** and drag it up to just after Boss-Extrude1. This will move the design back to just after the sprocket blank was created.

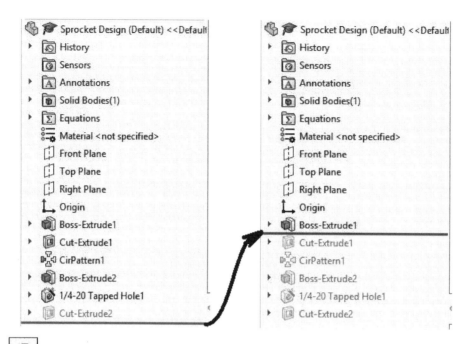

66. Use the Chamfer tool to add a 0.03 x 45° chamfer to both edges of the sprocket blank.

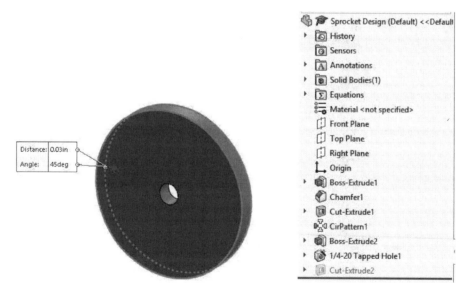

67. Select the blue line after the chamfer and drag it to the end of the **Feature Manager Design Tree** list. Each of the sprocket teeth now has a .03x45° chamfer.

68. Right-click on Equations in the **Feature Manager Design Tree**, then pick **Manage Equations…**
69. Change global variables, **Teeth** to **24**, **ChainNo** to **80**, **ShaftDia** to **1.0,** and **HubDia** to **2.5**, and then pick **OK.**

70. Right-click on Equations in the **Feature Manager Design Tree**, then pick **Manage Equations…**
71. Change global variables, **Teeth** to **15**, **ChainNo** to **60**, **ShaftDia** to **1.0,** and **HubDia** to **0.5**, and then pick **OK.**

72. Assign **Plain Carbon Steel** as the material.

73. 🖫 Save the part.

74. **File> Close**

Parametric Modeling Using Equations Problems

7-1   A series of right **"Cone Frustums"** needs to be designed to hold any specified volume given by the user. The parameters are the volume, the height, and the inside angle of the cone frustum will be specified. An angle of 90 degrees is not permitted although angles greater than 90 degrees are allowed. Assign Plain Carbon Steel as the material. Set the volume to 1000 in^3, the Height to 6 inches, and the Angle to 60 degrees. Use **Tools>Evaluate> Mass Properties…** to verify that the volume is 1000 in^3. Verify that your design works for any volume, height, and angle except for an angle of 90 degrees.

Name	Value / Equation	Evaluates to	
⊟ **Global Variables**			
"Volume"	= 1000	1000.000	
"Height"	= 6.00	6.000	
"Angle"	= 60	60.000	
"hTa"	= "Height" / Tan ("Angle")	3.464	
"BaseDia"	= (3 * "hTa" + sqr (36 * "Volume" / (pi * "Height") - 3 * "hTa" ^ 2)	17.893	

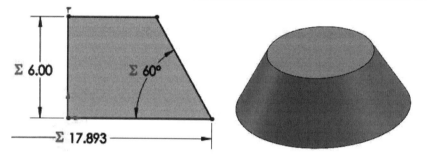

7-2   Design a **Flathead Screw** given the BasicSize, ThdsperInch, and Length such as ¼-20 x 1.00 inches. The HeadDia = 1.8374*BasicSize + 0.0196, the SlotDepth = (HeadDia-BasicSize)/3.5, the SlotWidth = 0.15*BasicSize + 0.035, and the chamfer on the end = 1/ThdsperInch. Material = Plain Carbon Steel.

   Verify your design works for ¼-28 x 1.50", 3/8-24 x 2.75", and ¾-10 x 3.00"

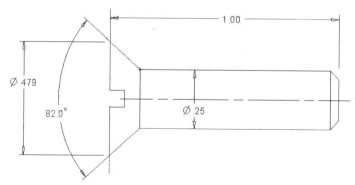

7-3   Design a commercial **"Mixing Bowl"** made from 0.035-inch thick AISI 321 annealed stainless steel. The parameters are the **Height** of the bowl (initially at 3.00 inches) and the approximate **Volume** (initially at 30 in³).

TopDia = sqr((Volume*6/(pi*Height)-Height^2)/0.9375)
BaseDia = TopDia/2
B=(0.75*TopDia^2+4*Height^2)
BaseHeight=(-B+sqr(B^2+4*(Height*TopDia)^2))/(Height*8)
Radius = (BaseDia^2+4*BaseHeight^2)/(8*BaseHeight)
BottomLocate = Radius-BaseHeight

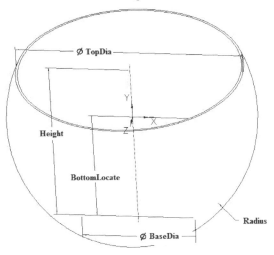

For a **Volume** of **30 in³** and **Height** of **3.00** inches, the following dimensions apply driven by the equations above. Note the Radius is a driven dimension in the sketch.

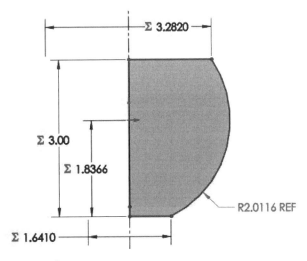

What if the **Volume** is **50** in³ and the **Height** is **4.00** inches?

Once you have verified that the total volume is correct, use the Shell icon to make the bowl hollow. Select the Shell icon, then pick the top surface of the bowl. Finally, set the thickness equal to 0.035 inches, then save the part.

7-4 Design a symmetrical **"Swivel Base"** made from Plain Carbon Steel. The diameter of the two mounting holes, **HoleDia** = 0.75 inches. The distance between the mounting holes, **Distance** = 4.00 inches. The thickness of the part, **Depth** = 0.25 inches. **HoleDia, Distance**, and **Depth** should be defined as Global variables. All the other dimensions are calculated based on these three global variables.

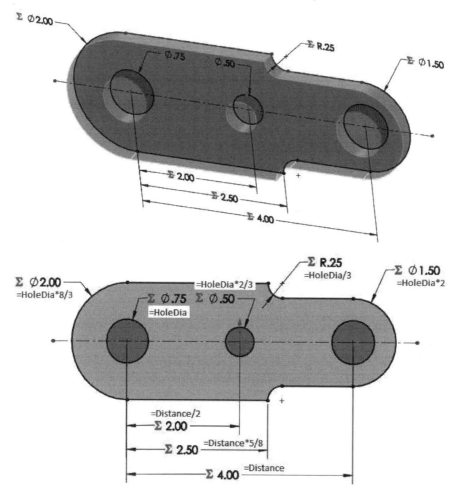

Once you have finished your design, change the three global variables. Set **HoleDia** to 1.00 inches, the **Distance** to 7.00 inches, and the **Depth** to 0.75 inches. Your design should look similar to the figure below.

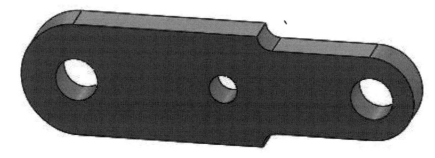

7-5 Design the **"Hinge Guide"** made for 1020 Steel. Base the design on the **HoleDia** = 0.75 inches, the **Distance** = 2.50 inches to the mounting hole, and the **Depth** = 2.00 inches of the part. The mounting hole diameter is **½ of the HoleDia** and a fixed ½-inch round is added at the junction of the flat portion and the round portion. Create the rest of the necessary equations to reflect the values shown below.

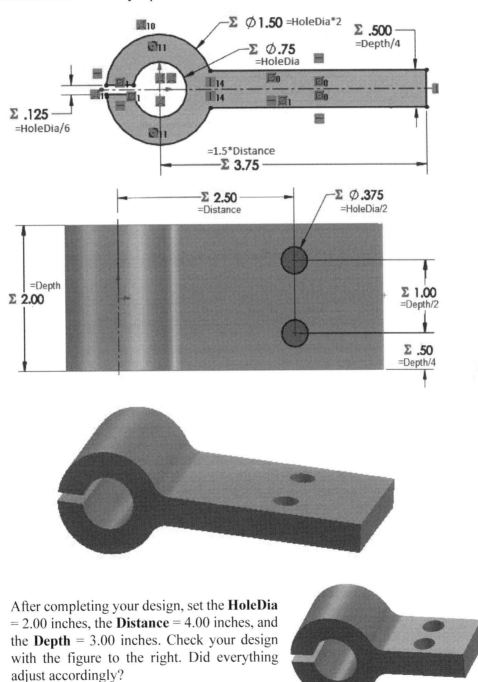

After completing your design, set the **HoleDia** = 2.00 inches, the **Distance** = 4.00 inches, and the **Depth** = 3.00 inches. Check your design with the figure to the right. Did everything adjust accordingly?

7-6   Design a **"Custom Nut"** made from brass. Define a global variable, **FlatDistance** =
        2.00 inches, which is the distance across the flats of the hex nut, and a global variable,
        Depth = 0.500 inches, which is the thickness of the nut. If the distance across the
        flats is less than 1.00 inches, then a round hole is placed in the center, otherwise a
        hex-shaped hole is placed in the center. There is a **.06-inch*Depth** thick boss on both
        sides of the hex nut, and the boss diameter is **FlatDistance*7/8.**

        For the figures shown below the distance across the flats is set at 2.00 inches.

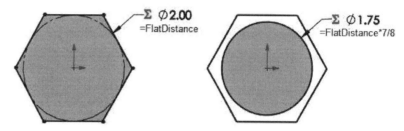

        The hex-shaped hole in the center is **FlatDistance*3/8**. The round hole in the
        center is **FlatDistance*7/16**.

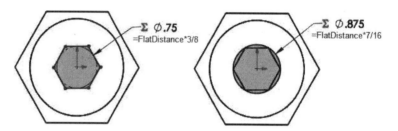

        If the distance across the flats is less or equal to 1.00-inch, then the center hole
        is used. If the center hole is greater than 1.00-inch the hex-shaped hole is used.

⊟ Features	
"Cut-Extrude1"	= iif ("FlatDistance" > 1.00 , "unsuppressed" , "Suppressed")
"Cut-Extrude2"	= iif ("FlatDistance" > 1.00 , "Suppressed" , "Unsuppressed")

        Verify that you have met the design intent by changing the distance across the
        flats to 1.00 inches. Set the Depth to .75 inches to verify that the thickness of
        the two bosses changes.

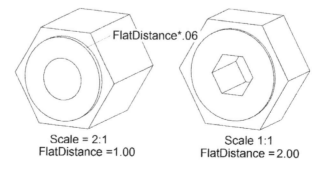

Scale = 2:1   Scale 1:1
FlatDistance =1.00   FlatDistance =2.00

Chapter 8.   Assemblies and Subassemblies

Introduction

Typically individual parts are not very useful, but rather how they fit together with other parts is important. The relationship between the parts will be discussed next. The best way to assemble a machine or structure in SOLIDWORKS is the same way it would be assembled in the real world. A large assembly should be made up of subassemblies and joining parts. Subassemblies (or groups of parts) are treated the same way as individual parts in SOLIDWORKS. Placement constraints create parent/child relationships that allow us to capture the design intent. A component that is placed on another part becomes the child of the already existing assembled parts.

The mating constraints are Coincident, Parallel, Perpendicular, Tangent, Concentric, Lock, Distance, and Angle. Initially, each part has six degrees of freedom, three rotations, and three translations. Each mating constraint removes one or more degrees of freedom. The movement of a fully constrained part is restricted in all directions.

The standard mating constraints are listed below.

Coincident – positions selected faces, planes, or edges so they share the same region, or two vertices are positioned so they touch.

Parallel – places the selected surfaces so they lie in the same direction and remain a constant distance apart.

Perpendicular – places the selected surfaces at a 90-degree angle relative to each other.

Tangent – places the selected surfaces tangent to one another. At least one surface must be a cylindrical, conical, or spherical surface.

Concentric – places the selected surface or edge so they share the same center point.

Lock – removes all six degrees of freedom. This should be used for the first part of an assembly or subassembly.

Distance – places the selected surface, edge, or point a specified distance from the second part's surface, edge, or point.

Angle – places the selected surface at a specified angle from the other part's surface.

Let's assume that a spur gear is to be pressed onto a step shaft. The step shaft is brought into the assembly first then a Lock mating constraint is applied which removes all six degrees of freedom. It doesn't matter whether the step shaft rotates in the final assembly, but for this subassembly, it is the first part so it is locked into place.

The spur gear is brought into the assembly next. It originally had six degrees of freedom, three rotations, and three translations. If we make the hole in the spur gear concentric with the shaft's outer surface, we remove four of the degrees of freedom. After this constraint is applied the spur gear is restricted to one rotation about the shaft's axis and one translation along the shaft.

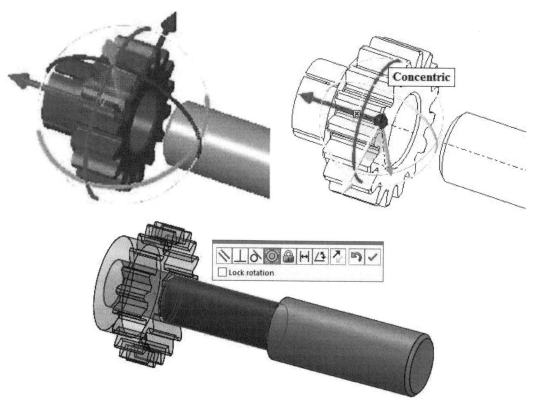

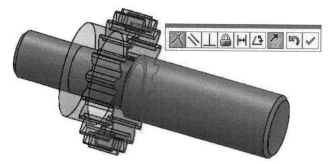

Next, we will slide the spur gear onto the shaft and butt it up against the diameter change in the step shaft using a coincident mating constraint.

The spur gear can no longer translate but it can still rotate about the step shaft's axis. If you pick the edge of one of the gear teeth and move the mouse cursor you can make the spur gear rotate about the step shaft's axis. A third mating constraint is needed. This time we will make the Front plane of the step shaft and a corresponding plane of the spur gear coincident. Now all six degrees of freedom have been removed from the spur gear. The spur gear and the step shaft will now move as one unit.

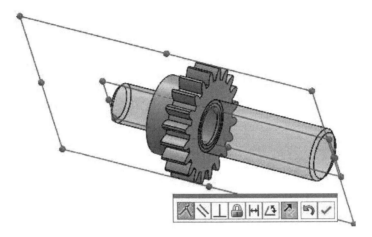

In summary, the spur gear is connected to the step shaft using one concentric and two coincident mating constraints. The order is based on which part was selected first in the mating process.

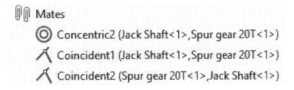

Assembly Practice

This practice session uses several of the different mating connections to put together a simple swinging link assembly.

Design Intent – Assemble the **"Swinging Link Assembly"**. The link must be able to rotate about the clevis pin and be held on by the retaining ring.

Before beginning, be sure the following four parts are in your working directory. You can create these parts, or get them from your instructor.

The **"Base"** is made from plain carbon steel and is shown below.

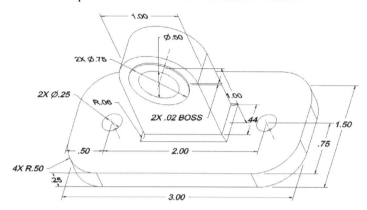

The **"Link"** is made from plain carbon steel and is shown below.

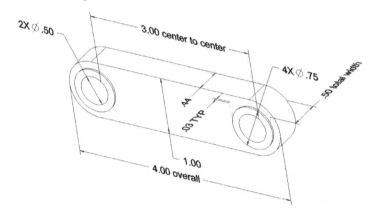

The **"Clevis Pin"** is made from AISI 1020 cold-rolled steel and is shown below.

Clip Included	A	B	C	E	F	G	H
No	0.5000	0.625	0.110	0.046	0.388	1.218	1.005

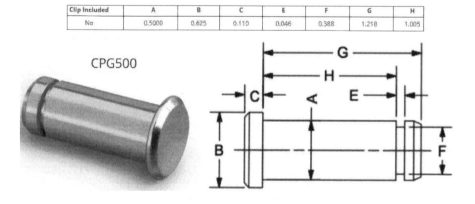

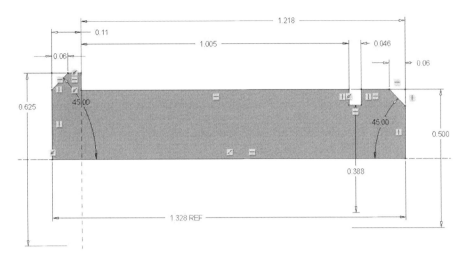

The E-style **"Retaining Ring"** (B27.1 – NA3-50) is made from AISI 1045 cold-drawn steel and is shown below. If the Design Library is available, perform the following steps to get the retaining ring, otherwise, get it from your instructor.

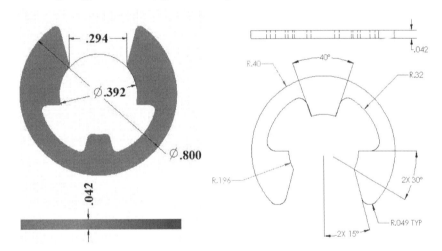

1. Start SOLIDWORKS if necessary.

2. Pick the Design Library icon.

3. **Toolbox** Pick the Toolbox icon.

4. **ANSI Inch** Pick the ANSI Inch folder icon.

5. **Retaining Rings** Pick the Retaining Rings folder icon.

6.   External   Pick the External folder icon.

E-TYPE-NA3-ANSI
7.   B27.1   Right-click on the E-TYPE-NA2-ANSI-B27.1 icon to bring up a
pop-up menu.
8. Pick Create Part from this pop-up menu.
9. Select ½ for the shaft size and verify the thickness is 0.042 inches. The part size
    should be NA3-50 needing a groove diameter of 0.396 and a groove width of 0.046
    inches. The default filename is "B27.1 – NA3–50".

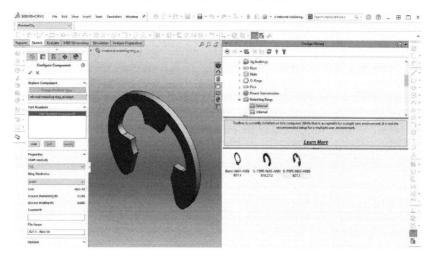

10. Pick the green checkmark to create the part.
11. Check anywhere in the graphics area to continue.
12. Assign AISI 1045 cold-drawn steel as its material.
13. Under File> pick **Properties…** For Property Name enter Material. Move to the
    Value/Text Expression area and enter "SW-Material". Pick OK.
14. Save the Retaining Ring in your working directory as **"Retaining Ring"**.
15. **File> Close**

Now let's begin assembling the swinging link assembly since all four parts are present.

1.   2024   Start SOLIDWORKS if necessary, then **File>New…**

2. Assembly In Advanced mode, pick the Template tab, then pick the Assembly icon
   and verify the units are IPS.

3. Select the **"Base"** from your working directory, then pick Open.
4. Move the Base to the center of the graphics area, then press the LMB. By default, the first part should have a mating constraint of Fixed.

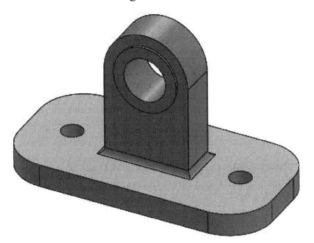

5. Pick the Insert Components icon in the upper left corner of the window.
6. Select the **"Clevis Pin"** from your working directory, then pick Open.
7. Move the Clevis Pin near the Base, then press the LMB.
8. Pick the Mate icon in the upper left corner of the window.
9. Pick the outside diameter of the Clevis Pin and the upper hole in the Base. The clevis pin should move into the Base's hole using the Concentric mating constraint. Check the Lock rotation box to prevent the Clevis Pin from rotating in the Base's hole, then pick the green checkmark.

10. It may look like the Clevis Pin is fixed in place, but it isn't. It can still slide in and out of the Base's hole. Pick the Clevis Pin, then try moving it. It still moves so we need another mating constraint.
11. Pick the flat mating surface of the Clevis Pin and the corresponding surface of the Base's boss. The Clevis pin should slide into the Base's hole and become fixed

using the Coincident mating constraint. The Clevis Pin no longer moves relative to the Base. Pick the green checkmark to accept the mating constraints, then pick the red X to exit from this mode. Reposition the subassembly in the graphics window.

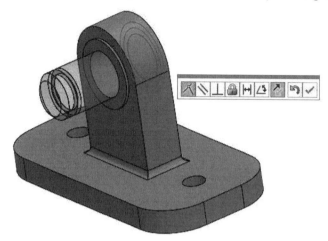

12. 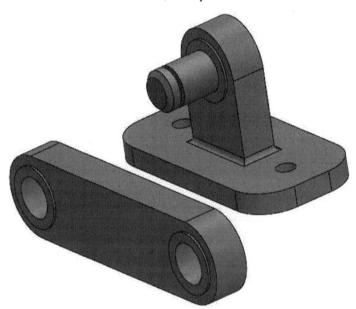 Select the Insert Components icon again.
13. Select the **"Link"** from your working directory, then pick Open.
14. Move the Link near the Clevis Pin, then press the LMB.

15. Pick the Mate icon in the upper left corner of the window.
16. Pick the left hole in the Link and the outside diameter of the Clevis Pin. The Link should move onto the Clevis Pin using the Concentric mating constraint. Do not check the Lock rotation box this time because we want the Link to rotate about the Clevis Pin. Pick the green checkmark.

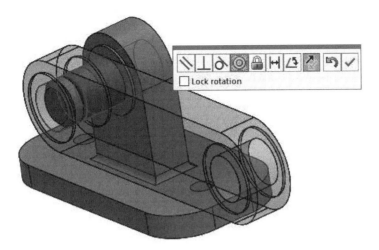

17. Pick the front boss's surface of the Base, then rotate the figure to select the back boss surface of the Link. The mating constraint should be Coincident.

18. 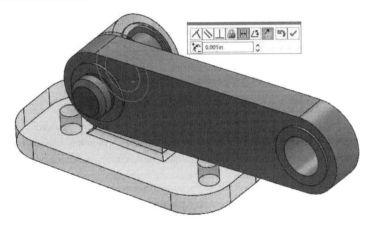 Instead of the Coincident constraint, we want to use the Distance mating constraint. Set the distance equal to 0.001 inches. Do this now.

19. Pick the green checkmark to accept the mating constraints, then pick the red X to exit from this mode.

20. If you pick an edge of the Link, then move the mouse cursor. The link should rotate about the Clevis Pin and be able to go through the Base since no restriction is present to prevent this.

21. Select the Insert Components icon again.

22. Select the **"Retaining Ring"** ("B27.1 – NA3–50") from your working directory, then pick Open.

23. Move the Retaining Ring near the Clevis Pin, then press the LMB.

24. Pick the Mate icon in the upper left corner of the window.

25. Pick an inside diameter portion of the Retaining Ring and the inner diameter of the groove in the Clevis pin. The Retaining Ring should move near the groove of the

Clevis Pin using the Concentric mating constraint. Do not check the Lock rotation box. Pick the green checkmark.

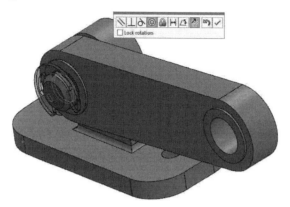

26. Zoom in, then pick the Retaining Ring and move it away from the Clevis Pin, then pick one side of the Retaining Ring and the corresponding side of the Clevis Pin's groove. The mating constraint should be Coincident. Pick the green checkmark to accept the mating constraints, then pick the red X to exit from this mode.

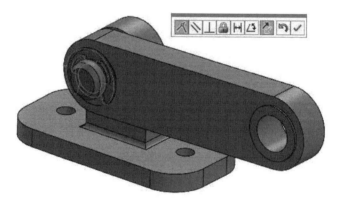

27. Pick the Link and make sure you can still rotate it about the Clevis Pin.

28. 💾 Save the assembly. Name it "**Swinging Link Assembly**".
29. **File> Print...** (if asked to)

Exploded View - In many cases, we want to show how the parts fit together. We can do this by creating an exploded assembly view and then saving the view.

1. [icon] With the Assembly tab still selected, pick the Exploded View icon at the top of the screen.

2. [icon] Pick the Configuration Manager icon at the top of the left window.

3. Pick the **Retaining Ring** (B27.1 – NA3–50), then move it away from the assembly. Pick the Done button when it appears far enough away.

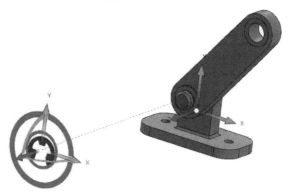

4. Pick the **Link**, then move it about half way between the Retaining Ring and the Base. Pick the Done button again.

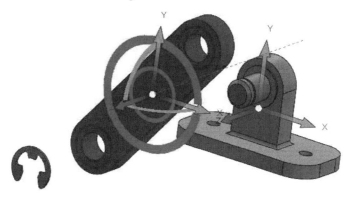

5. Pick the **Clevis Pin**, then move it back away from the Base. Pick the Done button.

6. When your exploded view looks similar to the following figure, pick the green checkmark.

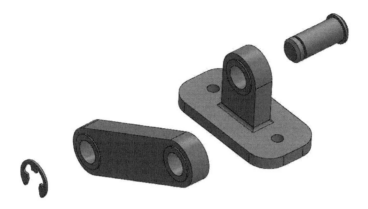

7. 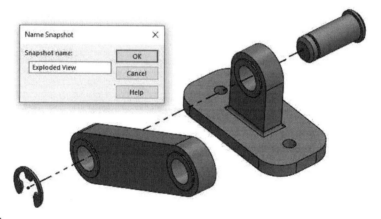 Pick the *Explode Line Sketch* icon. Pick a point on the end of the Clevis Pin, then the large hole in the Base, then the lower hole in the Link, and finally an inner surface of the Retaining Ring (B27.1 – NA3 – 50). Pick the green checkmark, then pick the red X.

8. Pick the Assembly tab again, then pick the *Explode Line Sketch* icon again to turn it off.

9. Pick the *Take Snapshot* icon. For the Snapshot name, enter "Exploded View" and pick OK.

10. Pick the Configuration Manager in the Feature Manager Design Tree, expand the Default list, then double-click on the Exploded View to collapse the exploded view.

11. Double-click on the Exploded View again to recreate the saved exploded view.

12. **File> Save**

13. **File> Print...** (if asked to)

14. **File> Close**

Assembly Exercise

Design Intent – Assemble a side-mounting **Take-Up Frame** made up of four different plain carbon steel parts shown below. The actual assembly will be welded.

Before you begin a new assembly you should consider how the individual parts fit together. In the picture below Part 2 is joined to Part 1, Part 3 is joined to Part 2, and Part 4 is joined to Part 1. The procedure of creating the parts, and then assembling them is called "Bottom-Up Design." The reverse procedure is called "Top-Down Design" where the parts are created within the assembly as needed. The two methods are not mutually exclusive, thus you can switch back and forth during the assembly process.

For this exercise, we will use the Bottom-Up Design process. This means that the parts are created and numbered, then assembled.

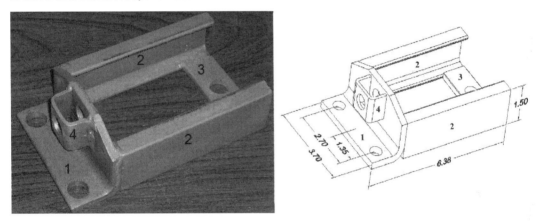

Part 1 is usually the main part of the assembly. We will number the **"Lower Plate"** made from plain carbon steel as part 1.

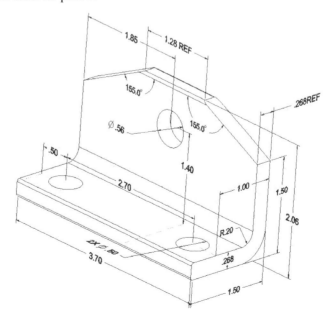

The **"Side Slide"** made from C15x14x488 C-channel plain carbon steel is part number 2.

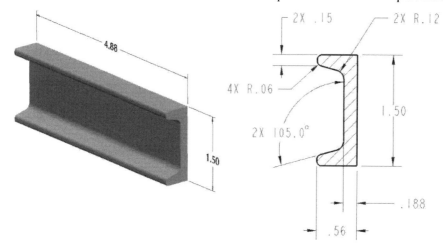

The **"Upper Plate"** made from ¼ inch by 1-inch plain carbon steel is part number 3.

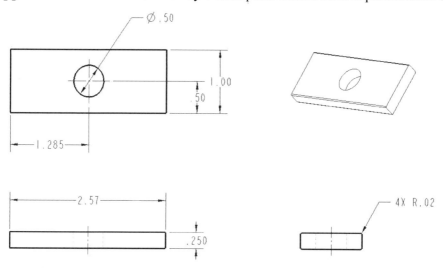

The **"Nut Holder"** is made from plain carbon steel bar stock bent in two places and is part number 4.

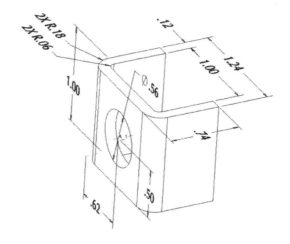

These four parts should be created in the IPS system and must be in the same working directory before continuing. If any of these four parts are missing, create them or get them from your instructor before continuing.

1. Start SOLIDWORKS if necessary, then **File>New...**

2. In Advanced mode, pick the Template tab, then pick the Assembly icon.
3. Select the **Lower Plate** from your working directory, then pick Open.
4. Move the Lower Plate to the center of the graphics area, then press the LMB. By default, the first part should have a mating constraint of Fixed.

5. Pick the Insert Components icon in the upper left corner of the window.
6. Select the **Side Slide** from your working directory, then pick Open.
7. Move the Side Slide near the Lower Plate, then press the LMB.

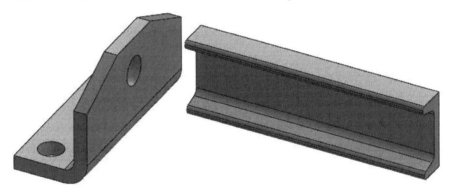

8. Pick the Mate icon in the upper left corner of the window.
9. Pick the left end of the Side Slider and the back surface of the Lower Plate. The Side Slide should move against the Lower Plate using the coincident mating constraint. Pick the green checkmark.

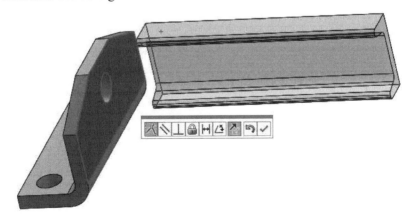

10. Pick the bottom surface of the Side Slide and the bottom surface of the Lower Plate. Again, coincident mating should be used. Pick the green checkmark.

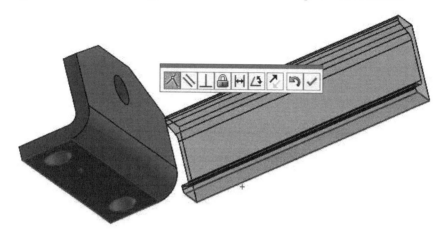

11. Pick the outside surface of the Side Slide and the outside surface of the Lower Plate. Again, coincident mating should be used. Pick the green checkmark.
12. After picking the green checkmark, pick the red X to exit from the Mating session.

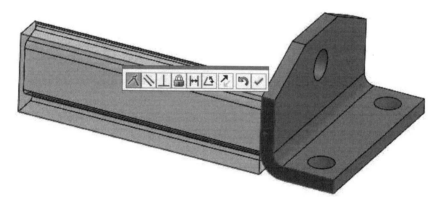

The Side Slide is now fixed to the Lower Plate and has no degrees of freedom.

13. Pick the Insert Components icon in the upper left corner of the window.
14. Select the **Side Slide** again from your working directory, then pick Open.

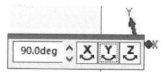

15. In the lower-left corner of the window, pick the Rotate about Y by 90° twice.
16. Move the Side Slide near the opposite side of the Lower Plate, then press the LMB.

17. Pick the Mate icon in the upper left corner of the window.

18. Pick the right end of the Side Slide and the back surface of the Lower Plate. The Side Slide should move against the Lower Plate using the coincident mating constraint. Pick the green checkmark.

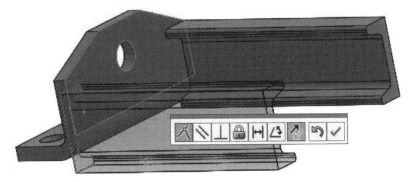

19. Pick the outside surface of the Side Slide and the outside surface of the Lower Plate. Again, coincident mating should be used. Pick the green checkmark.

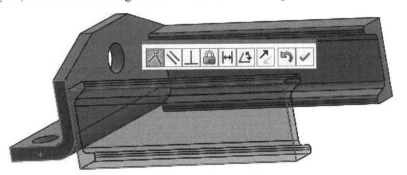

20. Pick the bottom surface of the Side Slide and the bottom surface of the Lower Plate. Again, coincident mating should be used. Pick the green checkmark.
21. After picking the green checkmark, pick the red X to exit from the Mating session.

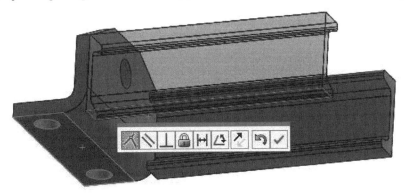

22. ⬚ Pick the Insert Components icon in the upper left corner of the window.
23. Select the **Upper Plate** from your working directory, then pick Open.
24. ⬚ In the lower-left corner of the window, pick the Rotate about Y by 90° once.
25. Move the Upper Plate near the Side Slide, then press the LMB.

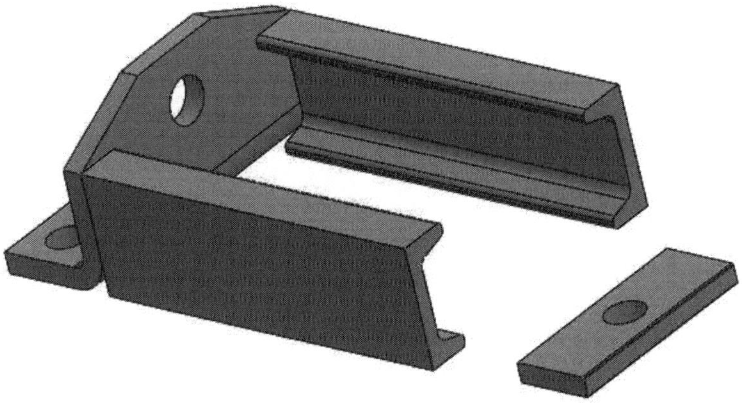

26. Pick the Mate icon in the upper left corner of the window.
27. Pick one end of the Upper Plate and the small flat surface of the Side Slide. The Upper Plate should move against the Side Slide using the coincident mating constraint. Pick the green checkmark.

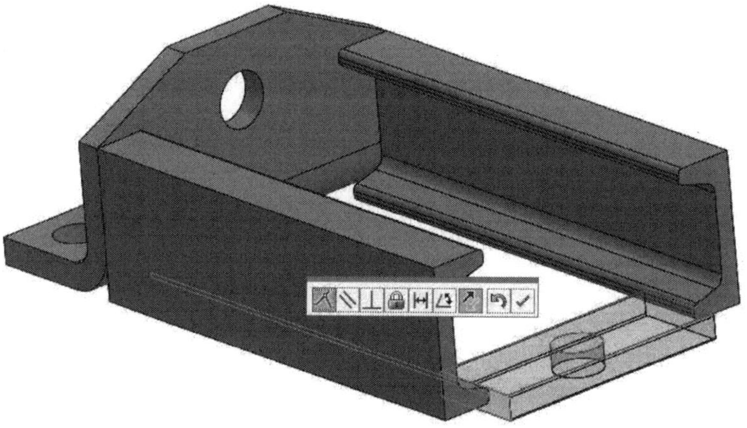

28. Pick the side of the Upper Plate and the end of the Side Slide. The Upper Plate should move flush with the Side Slide using the coincident mating constraint. Pick the green checkmark.

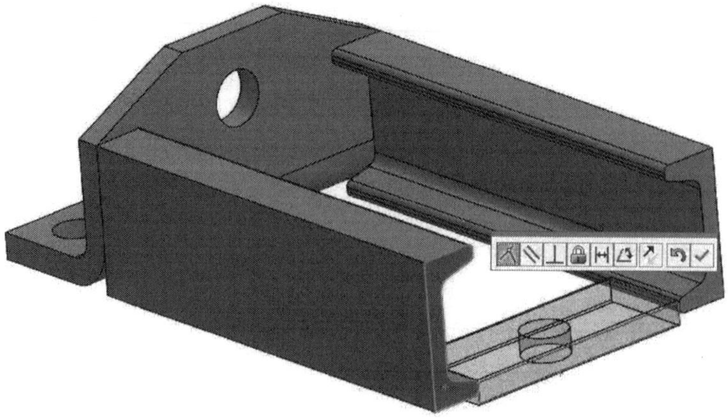

29. Pick the bottom of the Upper Plate and the bottom of the Side Slide. The Upper Plate should move up and be flush with the bottom of the Side Slide using the coincident mating constraint.

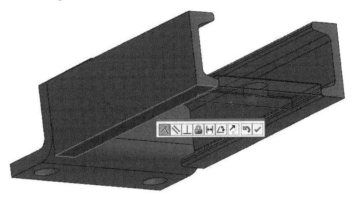

30. Pick the green checkmark, then pick the red X to exit from this Mating session.

31. Pick the Insert Components icon in the upper left corner of the window.

32. Select the **Nut Holder** from your working directory, then pick Open.

33. In the lower-left corner of the window, pick the Rotate about Y by 90° three times.

34. Move the Nut Holder near the Lower Plate, then press the LMB.

35. Pick the Mate icon in the upper left corner of the window.

36. Pick the hole in the Nut Holder and the hole in the Lower Plate. The Nut Holder should align itself with the Lower Plate using the Concentric mating constraint. Pick the green checkmark.

37. Pick the mating surface of the Nut Holder and the outer surface of the Lower Plate. A Coincident mating constraint is used for this. Pick the green checkmark.

38. Pick the top surface of the Nut Holder and the top surface of the Lower Plate. The parallel mating constraint should be used for this mating. Pick the green checkmark, then pick the red X to exit from the Mating session.

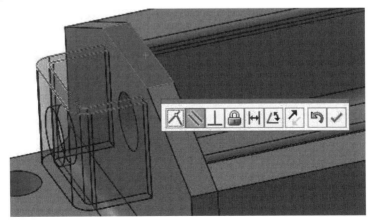

39. The **Take-Up Frame** is now completely assembled.

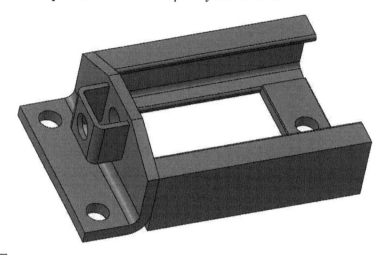

40. 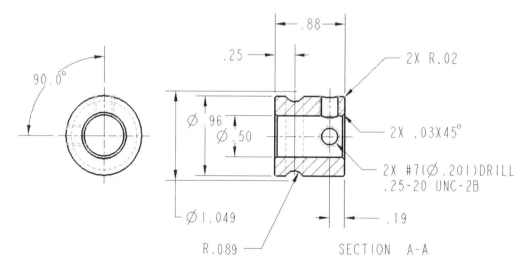 Save the assembly. Name it **"Take-Up Frame"**.
41. **File> Print...** (if asked to)
42. **File> Close**

You can stop at this point or go on with this exercise. If you choose to continue, be sure that the following four parts are located in your working directory. These parts should be created in the IPS unit system. In order to create this **Ball Bearing** assembly, all four parts must be in the same working directory. Be sure they are before continuing.

Part 1 is the **"Inner Race"** of the ball bearing made from AISI 4130 steel, normalized at 870°C.

Part 2 is the **"Outer Race"** of the ball bearing made from AISI 4130 steel, normalized at 870°C.

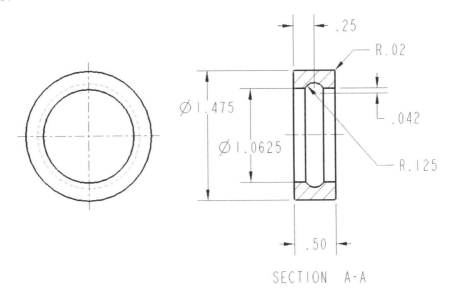

SECTION A-A

Part 3 is the 1/4 – 20 UNC **"Set Screw"** made from plain carbon steel.

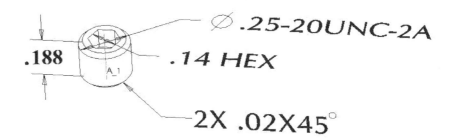

Part 4 is the **"Bearing Ball"** made from AISI 4130 steel, normalized at 870°C.

Design Intent – Assemble the **Ball Bearing** shown below from the four different steel parts shown above.

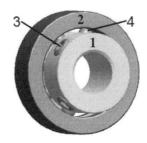

1. Start SOLIDWORKS if necessary, then **File>New...**

2. **Assembly** In Advanced mode, pick the Template tab, then pick the Assembly icon.

3. Select the **"Inner Race"** from your working directory, then pick Open.

4. Move the Inner Race to the center of the graphics area, then press the LMB. By default, the first part should have a mating constraint of Fixed.

5. Pick the arrow beside the View Orientation icon at the top of the screen, then pick the Right View from the pop-up menu to view the Inner Race from the right side.

6. Pick the Insert Components icon in the upper left corner of the window.

7. Select the **"Bearing Ball"** from your working directory, then pick Open.

8. Move the Bearing Ball above the Inner Race and above the ball track, then press the LMB.

9. Pick the Mate icon in the upper left corner of the window.

10. Pick the bottom of the bearing ball, then pick the bottom center of the ball track. The mating constraint should be Tangent, and the ball should be symmetrical in the ball track. Pick the green checkmark, then pick the red X to exit from this session.

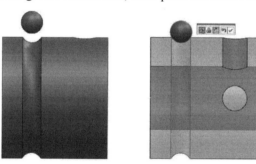

11. Pick the Circular Component Pattern icon from the top of the window.

12. For Direction 1, pick the center hole in the Inner Race. With the range set to 360 degrees, enter 10 for the quantity. Check the Equal spacing box. Pick are area in the Component to Pattern section, then select the Bearing Ball. Ten Bearing Balls should appear around the Inner Race. Pick the green checkmark.

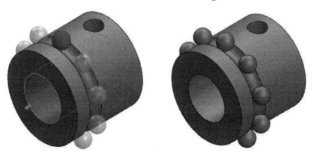

13.  Pick the Insert Components icon in the upper left corner of the window.

14. Select the **"Outer Race"** from your working directory, then pick Open.

15. Move the Outer Race near the Inner Race, then press the LMB.

16. Pick the Mate icon in the upper left corner of the window.

17. Pick the outer surface of the Outer Race and the outer surface of the Inner Race, then pick the green checkmark.

18. Pick the left flat surface of the Outer Race and the left flat surface of the Inner Race, then pick the green checkmark.

19. Pick the red X to exit from this session.

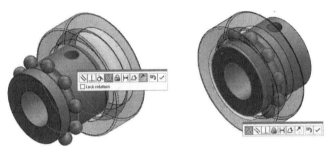

20. Pick the Insert Components icon in the upper left corner of the window.

21. Select the **"Set Screw"** from your working directory, then pick Open.

22. Move the Set Screw near one of the Inner Race screw holes, then press the LMB.

23. Pick the Mate icon in the upper left corner of the window.

24. Pick the *Mechanical tab*, then pick *Screw* from the list.

25. Pick the area below the Mate Selection area, then pick the Set Screw and the screw hole in the Inner Race. Enter 20 for the number of revolutions per inch. Be sure the Reverse box is not checked. (Pick the Mate alignment icon if the set screw is backward in the hole.)

26. Pick the green checkmark, then pick the red X to exit from this session.

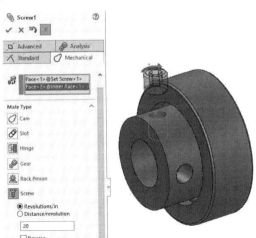

27. Pick one of the edges of the setscrew's hex slot, then move the mouse cursor around the Set Screw as though you were screwing it into the hole. Continue doing this until the top of the setscrew is near the surface of the ball bearing hub.

28. Repeat steps 20 through 27 to insert the second set screw in the second threaded hole of the ball bearing hub. When the second set screw is properly placed in the second hole, pick the green checkmark, then pick the red X to exit the session.

29. ▣ Save the assembly. Name it "**Ball Bearing**."
30. **File> Print...** (if asked to)
31. **File> Close**

Before you continue, be sure that the following parts are in your working directory. Parts TU250 and Grease Fitting should be created in the IPS system. Check with your instructor to see if these parts are available for your use without creating them.

Design Intent – Assemble the **Take-up Piece** by inserting the Ball Bearing assembly and the Grease Fitting into TU250.

Part **"TU250"** is made from plain carbon steel and is shown below.

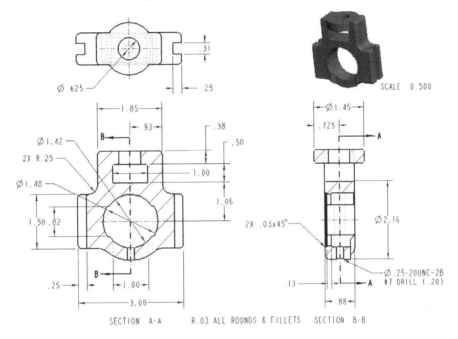

The **"Grease Fitting"** is made from plain carbon steel and is shown below.

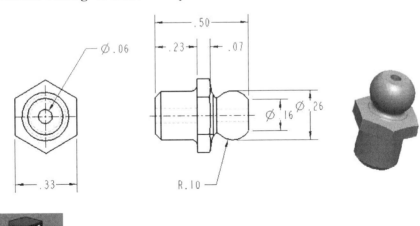

1.   Start SOLIDWORKS if necessary, then **File>New…**

2. In Advanced mode, pick the Template tab, then pick the Assembly icon.
3. Select **"TU250"** from your working directory, then pick Open.
4. Move TU250 to the center of the graphics area, then press the LMB. By default, the first part should have a mating constraint of Fixed.

5. Pick the Insert Components icon in the upper left corner of the window.
6. Select the **"Bearing Ball"** assembly from your working directory, then pick Open.
7. Move the Bearing Ball assembly near TU250, then press the LMB.

8. Pick the Mate icon in the upper left corner of the window.
9. Pick the outer surface of the bearing ball assembly, then pick the large hole in TU250. The mating constraint should be Concentric.
10. If the Ball Bearing assembly was inserted backward, pick the Mate alignment icon to rotate the Ball Bearing assembly by 180 degrees. Pick the green checkmark.
11. Pick the lower flat surface of TU250 and the side surface of the Ball Bearing assembly. The mating constraint should be Coincident.
12. Pick the green checkmark, then pick the red X to exit from this session.
13. Pick one of the edges of one of the setscrews in the Ball Bearing assembly. You should be able to make the Inner Race of the Ball Bearing assembly rotate. Pick anywhere in the graphics area to end this feature.

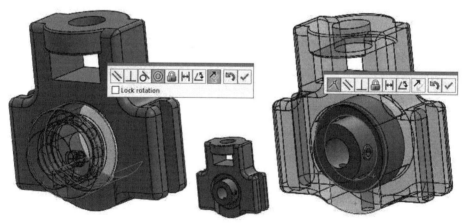

14. Tilt the TU250 so we can see the hole in the bottom portion of the part.

15. Pick the Insert Components icon in the upper left corner of the window.
16. Select the **"Grease Fitting"** from your working directory, then pick Open.
17. Move the Grease Fitting near TU250, then press the LMB.

18. Pick the Mate icon in the upper left corner of the window.

19. Pick the *Mechanical tab*, then pick *Screw* from the list.

20. Pick the area below the Mate Selection area. Pick the small screw hole at the bottom of TU250, then pick the threaded portion of the Grease Fitting. Enter 20 for the number of revolutions per inch. Be sure the Reverse box is not checked.

21. Pick the green checkmark, then pick the red X to exit from this session.

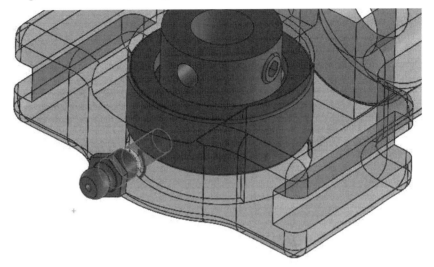

22. Pick a flat portion of the Grease Fitting, then move your mouse clockwise around the Grease Fitting like you are screwing it into the hole. Continue until the Grease Fitting is properly placed. Pick anywhere in the graphics area to end this feature.

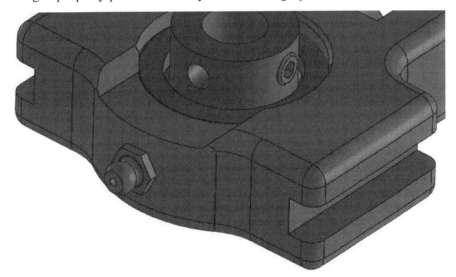

23. ▣ Save the assembly. Name it "**Take-up Piece**."

24. **File> Print…** (if asked to)

25. **File> Close**

Before you continue, be sure the following subassemblies and associated parts are in your working directory.

Design Intent – Assemble the entire **Take-up Assembly** using the Ball Bearing assembly, the Take-up Piece assembly, and the corresponding parts shown below. Turning the Nut on the end of the Threaded Rod should move the Take-up Piece along the Take-up Frame.

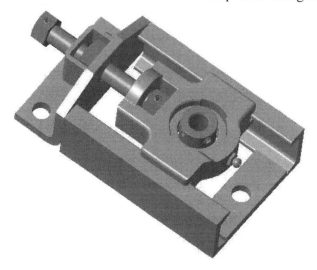

The **"Threaded Rod"** is made from plain carbon steel.

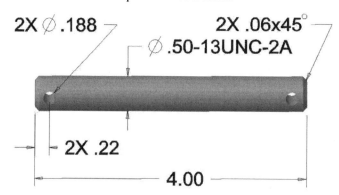

The **"Square Nut"** and the **"Hex Nut"** are made from plain carbon steel.

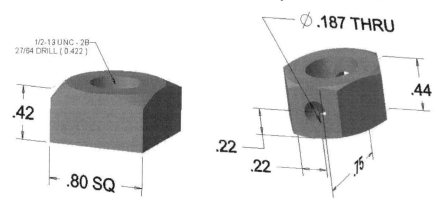

The **"Roll Pin"** is made from plain carbon steel.

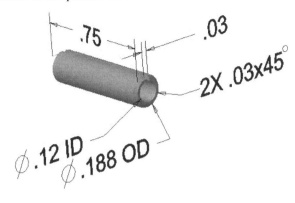

1.   Start SOLIDWORKS if necessary, then **File>New…**

2. In Advanced mode, pick the Template tab, then pick the Assembly icon.
3. Select **"Take-up Frame"** from your working directory, then pick Open.
4. Move the Take-up Frame to the center of the graphics area, then press the LMB. By default, the first part should have a mating constraint of Fixed.

5.   Pick the Insert Components icon.
6. Select the **"Square Nut"** from your working directory, then pick Open.

7.   Pick the Mate icon. Pick the hole in the Take-up Frame and the hole in the Nut. Make it a Concentric constraint. Flip the square nut if necessary using the Mate alignment option. Pick the green checkmark.

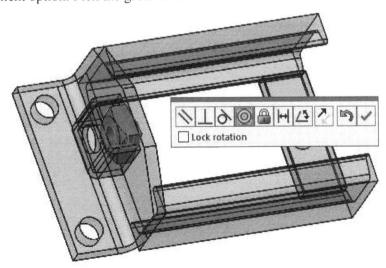

8. Make the top of the Take-up Frame parallel with the top of the Nut. Pick the green checkmark. Make the outer surface of the Lower Plate and the right side of the Nut Coincident. Pick the green checkmark. Pick the red X to exit from this session.

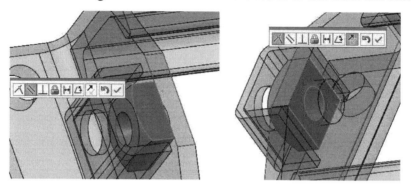

9. 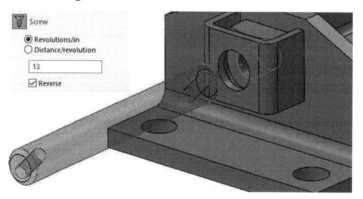 Save the assembly. Name it **"Take-up Assembly"**.
10. Pick the Insert Components icon.
11. Select the **"Threaded Rod"** from your working directory, then pick Open.
12. Pick the Mate icon. Select the *Mechanical tab*, then pick the *Screw* icon. Pick the Threaded Rod and the hole in the Square nut. Set the number of revolutions per inch to 13. Be sure the arrow on the threaded rod is pointed in the clockwise direction. Pick the green checkmark. Pick the red X to exit from this session.

13. Pick the hole at the end of the Threaded Rod, then move the cursor clockwise around the rod to verify that you are screwing the rod into the Square Nut.

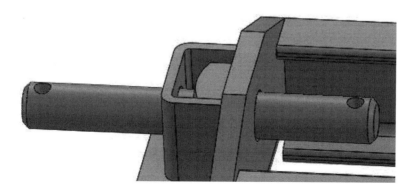

14. Pick the Insert Components icon.
15. Select the **"Hex Nut"** from your working directory, then pick Open.
16. Pick the Mate icon. Pick the hole in the HEX Nut and the outside diameter of the Threaded Rod outside the Take-Up Frame. The Mating constraint should be Concentric. Pick the green checkmark.
17. Pick the small hole in the Hex Nut and the small hole at the left end of the Threaded Rod. The Mating constraint should again be Concentric. Pick the green checkmark, then pick the red X.

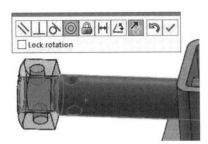 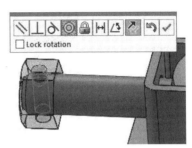

18. Pick the Insert Components icon.
19. Select the **"Roll Pin"** from your working directory, then pick Open.
20. Pick the Mate icon. Make the connection between the Roll Pin and the small hole in the left Hex Nut Concentric. Make the end of the Roll Pin and the flat surface of the left Hex Nut Coincident. Pick the green checkmark, then pick the red X.

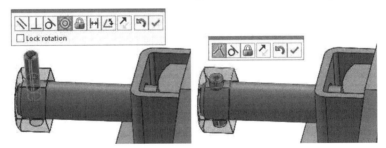

21. Pick the Insert Components icon.
22. Select the **"Hex Nut"** from your working directory again, then pick Open.
23. Pick the Mate icon. Pick the hole in the HEX Nut and the outside diameter of the Threaded Rod inside the Take-Up Frame. Make the Mating constraint Concentric.
24. Pick the small hole in the Hex Nut and the small hole at the right end of the Threaded Rod. The Mating constraint should again be Concentric. Pick the green checkmark, then pick the red X.

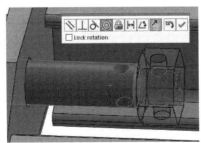 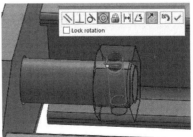

25. Pick the Insert Components icon.

26. Select the **"Roll Pin"** from your working directory again, then pick Open.

27. Pick the Mate icon. Make the connection between the Roll Pin and the small hole in the right Hex Nut Concentric. Make the end of the Roll Pin and the flat surface of the right Hex Nut Coincident. Pick the green checkmark, then pick the red X.

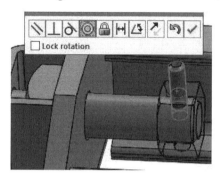 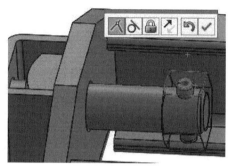

The Take-up Frame Assembly should be similar to the figure below.

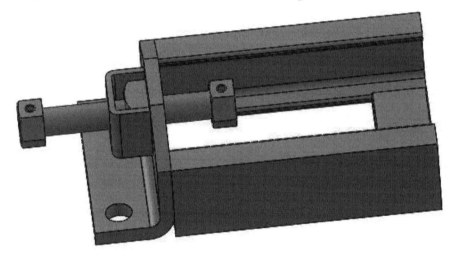

28. 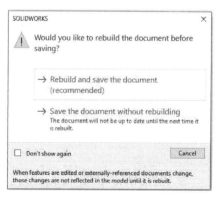 Save the assembly. When the message, "Would you like to rebuild the document before saving?" Comes up, pick the "Rebuild and save the document" option.

29. Rotate the Take-Up Assembly in the graphics area similar to the figure below.
30. Pick the Insert Components icon.
31. Select the **"Take-up Piece"** from your working directory, then pick Open.
32. Pick the Mate icon. Pick the hole in the Take-up Piece and the outside diameter of the Threaded Rod. The Mating constraint should be Concentric.

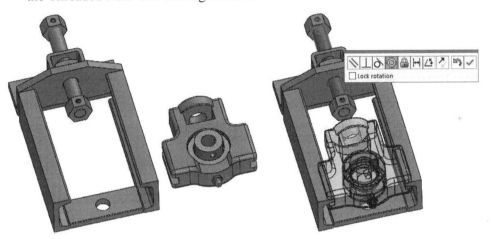

1. Add a Parallel constraint to the upper surface of the cutout of the Take-up Piece and the top surface of the Take-up Frame.
2. Make the inside surface of the Take-up Piece and the flat end of the Hex Nut a Coincident constraint.

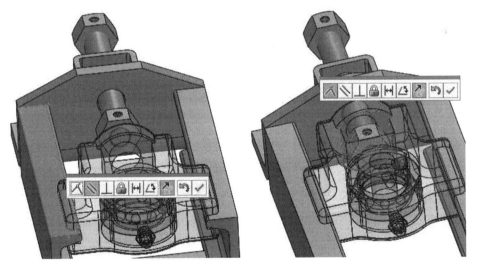

This completes the assembly of the Take-up Assembly.

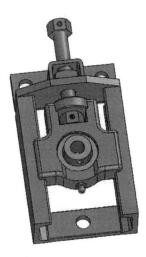

3. Pick the top surface of the Take-up Piece, then slide it along the Take-up Frame. The Threaded Rod and all its components should rotate as you slide it.

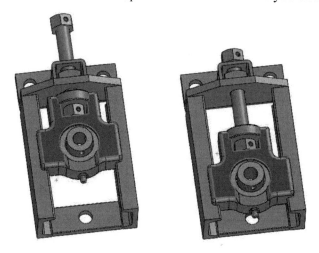

4. 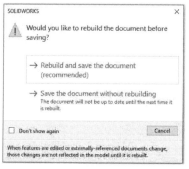Save the assembly. When the message, "Would you like to rebuild the document before saving?" appears, pick the "Rebuild and save the document" option.

5. **File> Print…** (if asked to)
6. **File> Close**

Assembly Problems

8-1   Assemble the pulley on the short step shaft shown below. Use the square key to allow the pulley to transmit torque to the shaft. The shaft is made from 1020 CR steel. The Square Key is made from 1010 HR bar steel. The 4.5-inch diameter V-belt pulley and Set Screw-Dogpt are made from plain carbon steel. You can create these parts or get them from your instructor. Start with the step shaft, then add the key and the pulley. Insert the set screw into the threaded hole of the pulley using the screw mating constraint. Name the assembly **"Step-Shaft Assembly"**.

The **"Short Step Shaft"** is made from 1020 Cold Rolled steel.

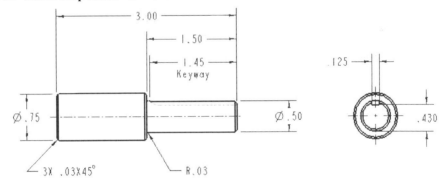

The **"Square Key"** is made from 1010 Hot Rolled Bar steel.

.125 SQUARE   1.25

The **"Setscrew-Dogpt"** is made from plain carbon steel.

—1/4-20UNC-2A

.50

The **"V-belt Pulley-4.5"** is made from plain carbon steel.

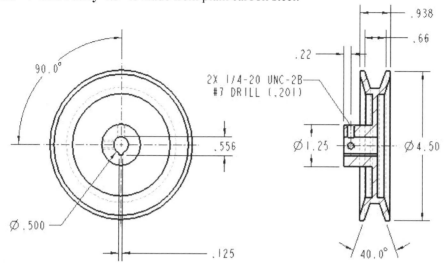

8-2   Assemble the **"Drill Jig Assembly"** from the pieces shown below. Start with the drill jig base. **Hint:** For the flathead screws in the end brackets, use the screw mating constraint, then make the top of the flathead screw even with the top surface of the end bracket. You can create the parts or get them from your instructor.

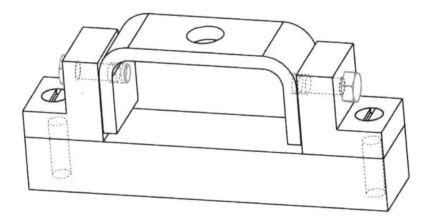

The **"Drill Jig Base"** is made from plain carbon steel.

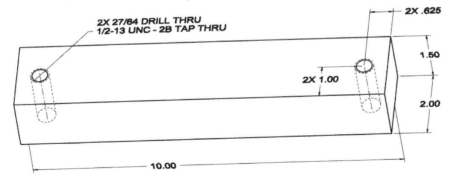

The **"End Bracket"** is made from plain carbon steel.

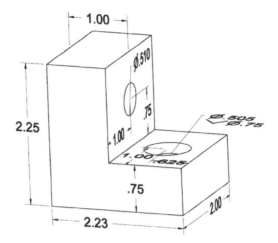

The ½-13 UNC x 1.25-inch **"Drill Jig Flathead Screw"** and the ½-13 UNC x 1.5-inch **"Drill Jig Capscrew"** are made from plain carbon steel.

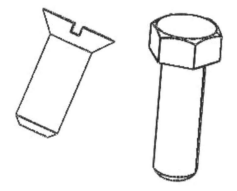

The **"Center Plate"** is made from plain carbon steel.

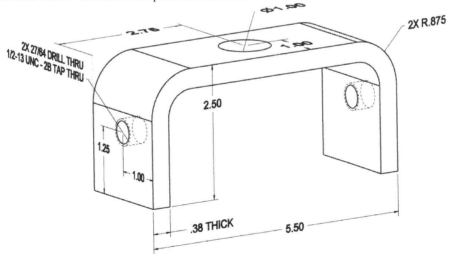

8-3   Create the "**Injector Mold Assembly**" shown below using the parts shown. You can create the parts or get them from your instructor. When making the assembly start with the Lower Die, then add the Dowel Pins, and finally add the Upper Die.

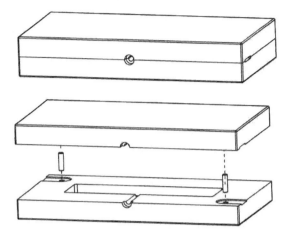

The **"Lower Die"** is made from plain carbon steel. The two cutouts are .03 inches deep. Use a #30 drill for the .130 diameter holes that are .25 inches deep.

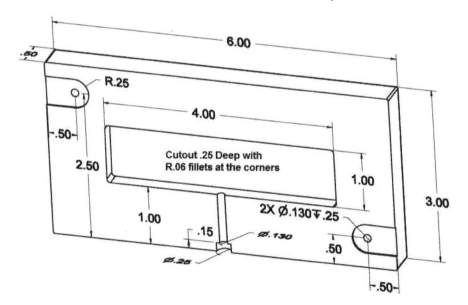

The **"Dowel Pin"** is made from plain carbon steel. There is a .01-inch x 45° chamfer on both ends. When inserting the Dowel Pin in the Lower Die, use a distance of 0.25 between the top of the Dowel Pin and the surface of the cutout.

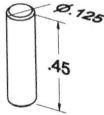

The **"Upper Die"** is made from plain carbon steel and is a mirror image of the Lower Plate. You can make a copy of the Lower Die, then move the two cutouts to create the Upper Die.

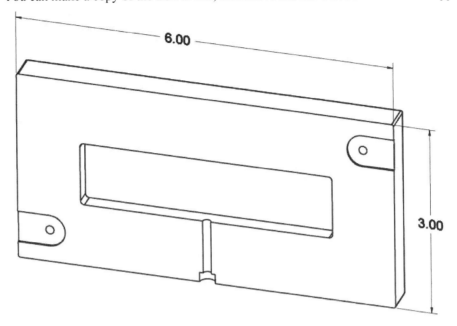

8-4   Assemble the **"Adjustable Shaft Support"** shown below using the parts that follow. You can create these parts or get them from your instructor. Start with the Base Support, then add the Vertical Shaft, etc.

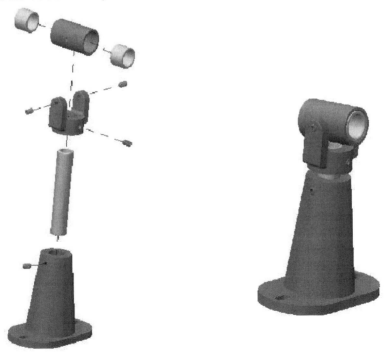

The **"Base Support"** is made from plain carbon steel.

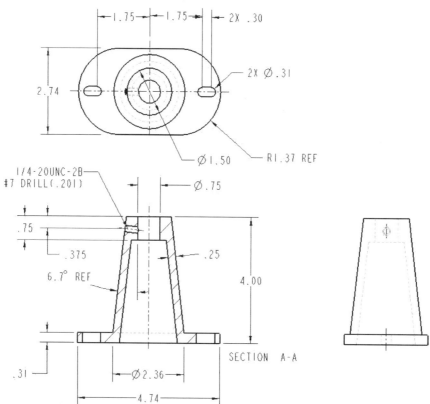

The **"Vertical Shaft"** is made from plain carbon steel. Install the Vertical Shaft so that it protrudes 2 inches above the top of the Base Support.

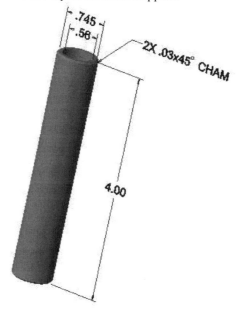

The **"Yoke"** is made from plain carbon steel.

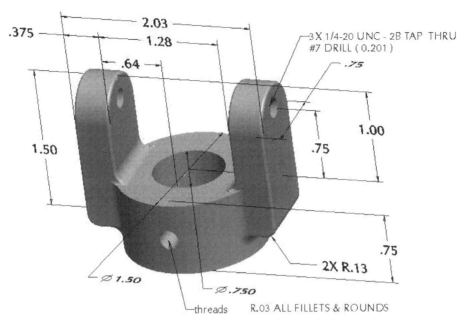

The **"Bushing Housing"** is made from plain carbon steel.

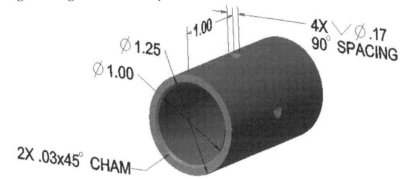

The **"Bushing"** is made from Tin Bearing Bronze, and the ¼-20 UNC **"Setscrew-Dogpt"** is made from plain carbon steel. Be sure to use the Screw mating constraint for the four Setscrews.

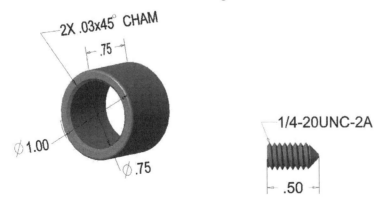

After the assembly is complete, add one more mating constraint. Force the Bushing Housing to be at a 9-degree angle relative to the horizontal.

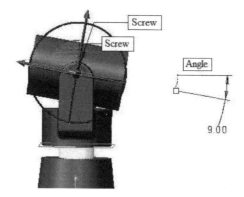

8-5 Create an exploded assembly view of the **"Injector Mold Assembly"** with exploded lines similar to the figure below.

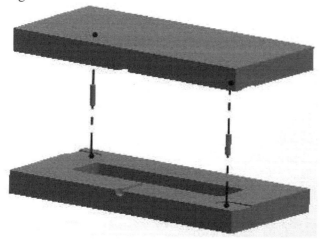

8-6 Create an exploded assembly view of the **"Step Shaft Assembly"**. Add exploded lines to show where the parts fit relative to each other.

8-7   Create an exploded assembly view of the **"Take-up Frame"**. Add exploded lines to show where the parts fit relative to each other.

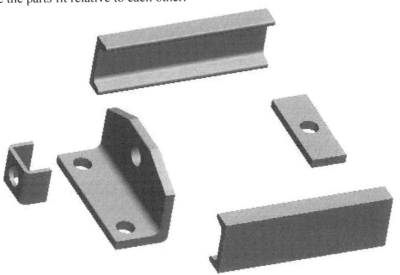

8-8   Create an exploded assembly view of the **"Take-up Piece"**. Add exploded lines to show where the parts fit relative to each other.

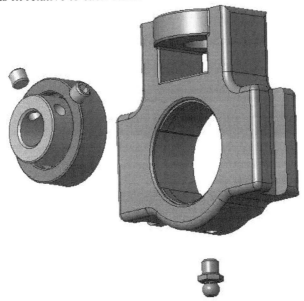

8-9 Assemble the **"Door Hanger Assembly"** shown below using the parts that follow. You can create these parts or get them from your instructor. Start with the Wooden Door, then add the Roller Bracket, etc.

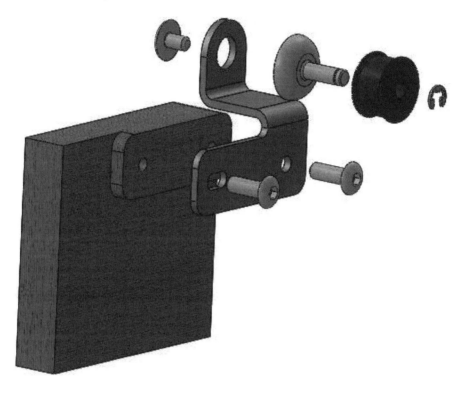

The **"Wooden Door"** is made from Oak.

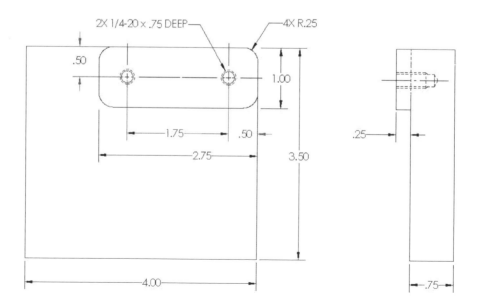

The **"Roller Bracket"** is made from AISI 1010 hot rolled bar steel. (See problem 10.6 for further details.)

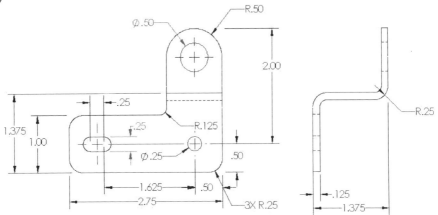

The ¼-20UNC-2A x ¾ **"Round Head Screw"** is made from AISI 316 annealed stainless steel bar.

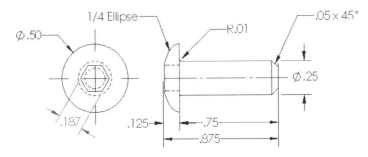

The **"Roller Stud"** is made from 201 annealed stainless steel.

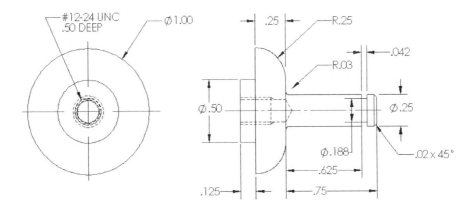

The #12-24 UNC-2A x 3/8 **"Round Head Screw-short"** is made from AISI 316 annealed stainless steel bar.

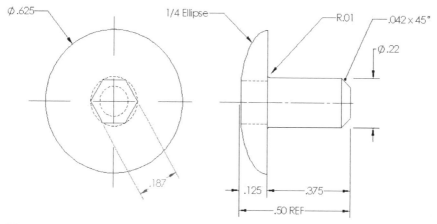

The **"Rubber Roller"** is made from butyl rubber.

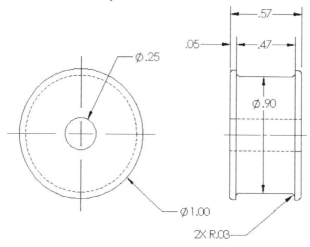

The **"Clip Pin"** is made from AISI 316 stainless steel sheet.

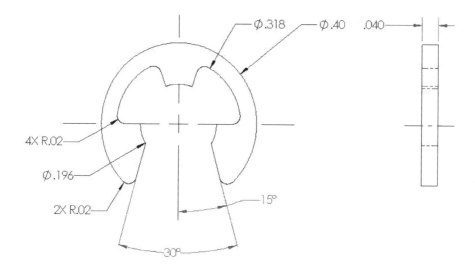

8-10 Assemble the **"Drilling Jig Assembly"** from the parts shown. You can create these parts or get them from your instructor.

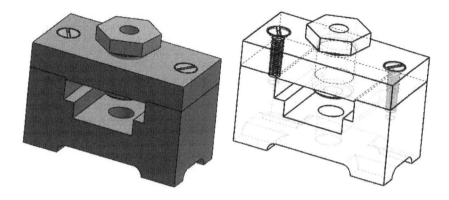

The **"Drilling Jig Base"** is made from plain carbon steel.

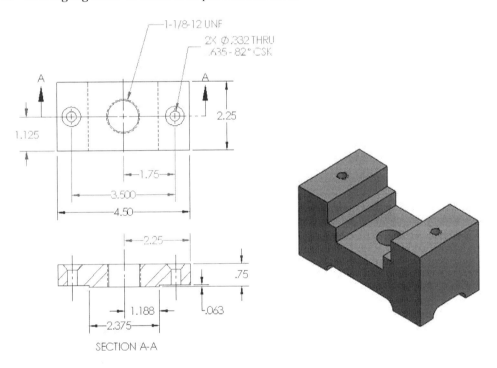

The 5/16-18 UNC-2A **"Drilling Jig Flathead Screw"** is made from AISI 1010 hot-rolled steel.

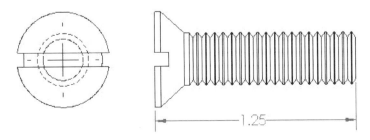

The **"Drilling Jig Plate"** is made from plain carbon steel.

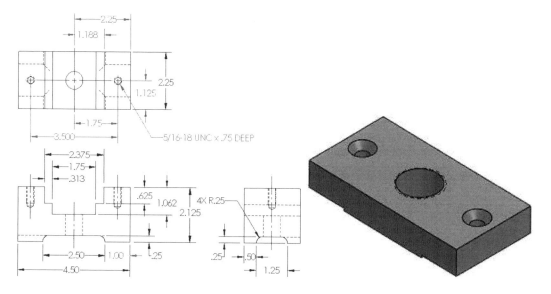

The 1.125-12 UNF-2A **"Drilling Jig Clamp Bushing"** is made from 1060 Alloy steel.

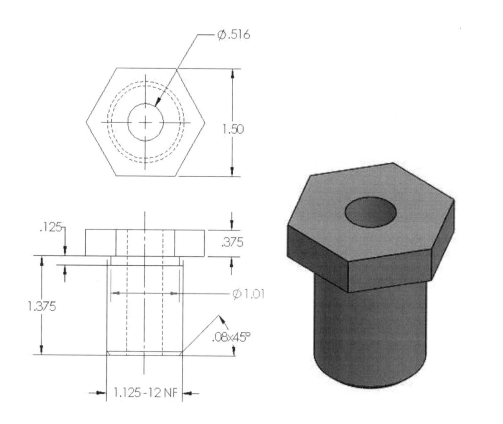

Chapter 9.   Assembly Drawing

Introduction

A set of working drawings includes detailed drawings of each part and an assembly drawing shows how the parts fit together. An assembly drawing is shown below.

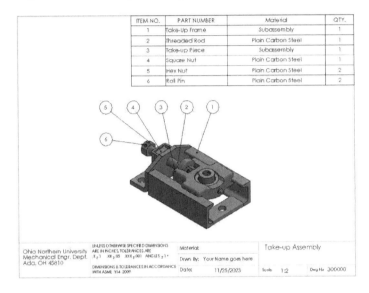

ITEM NO.	PART NUMBER	Material	QTY.
1	Take-Up Frame	Subassembly	1
2	Threaded Rod	Plain Carbon Steel	1
3	Take-up Piece	Subassembly	1
4	Square Nut	Plain Carbon Steel	1
5	Hex Nut	Plain Carbon Steel	2
6	Roll Pin	Plain Carbon Steel	2

A subassembly drawing shows only a portion of a larger assembly as shown below.

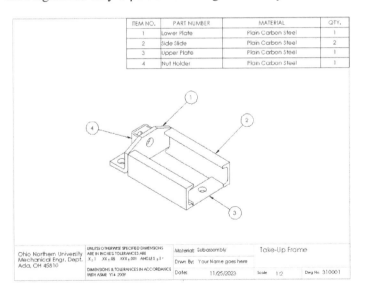

ITEM NO.	PART NUMBER	MATERIAL	QTY.
1	Lower Plate	Plain Carbon Steel	1
2	Side Slide	Plain Carbon Steel	2
3	Upper Plate	Plain Carbon Steel	1
4	Nut Holder	Plain Carbon Steel	1

The views shown in an assembly or subassembly drawing must show how the parts fit together and/or suggest how the device functions. Sometimes an exploded assembly view is used to show how the parts fit together.

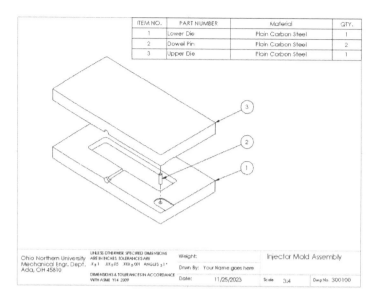

ITEM NO.	PART NUMBER	Material	QTY.
1	Lower Die	Plain Carbon Steel	1
2	Dowel Pin	Plain Carbon Steel	2
3	Upper Die	Plain Carbon Steel	1

Assembly or subassembly drawings can be made up of two orthographic views if this better shows how the parts fit together, as seen below.

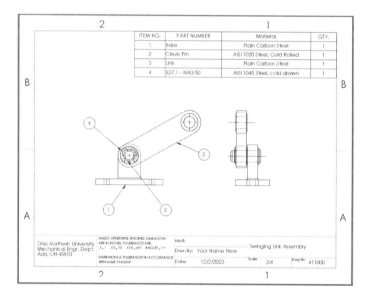

ITEM NO.	PART NUMBER	Material	QTY.
1	Base	Plain Carbon Steel	1
2	Clevis Pin	AISI 1020 Steel, Cold Rolled	1
3	Link	Plain Carbon Steel	1
4	B27.1 - NA3-50	AISI 1045 Steel, cold drawn	1

Since assembly or subassembly drawings show how parts fit together, sectioning at least one view may make it easier to understand how the parts fit together. A full-section, half-section, or several removed sections is the easiest way to show this. When sectioning parts, be sure to cross-hatch each part differently so the observer can distinguish between the parts. Small sectioned parts should have their section lines close together. Relatively thin parts, such as gaskets or sheet metal, should be shaded solid or not at all. Often solid parts that fall in the cutting plane do not need to be sectioned. It is customary to show these parts not sectioned or "in the round." These parts include screws, bolts, nuts, keys, pins, ball or roller bearings, gear teeth, and spokes. Hidden lines are often left out of an assembly drawing since the relationship between parts is the assembly drawing's main purpose. Hidden lines may tend to confuse the desired detail.

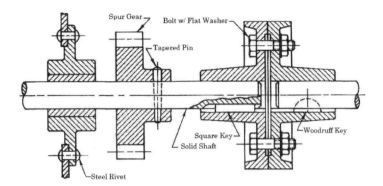

The purpose of a detailed drawing is to describe the size or shape of any part. Dimensions are shown on the detailed drawings; thus they are not included on the assembly drawing. A dimension can be shown on an assembly drawing if it indicates the amount of travel allowed by a given part or the overall height or width of the complete assembly. Keep this distinction in mind.

Parts are identified by placing a number inside a circle or balloon near the part with a leader line attaching it to the part. The circled numbers should be placed on the drawing in a clockwise or counter-clockwise fashion. The balloon's leader lines should never cross each other. Leader lines close to each other should be approximately parallel.

A parts list, sometimes called a bill of materials, should include the part numbers, a description of the part, its material or its classification, and the required number of pieces. The parts list is usually located in the upper right corner or lower right corner of the drawing, or on a separate sheet.

ITEM NO.	PART NUMBER	Material	QTY.
1	Take-Up Frame	Subassembly	1
2	Threaded Rod	Plain Carbon Steel	1
3	Take-up Piece	Subassembly	1
4	Square Nut	Plain Carbon Steel	1
5	Hex Nut	Plain Carbon Steel	2
6	Roll Pin	Plain Carbon Steel	2

The border and title block for an assembly drawing should be similar to that of an engineering detailed drawing. In the title block for a detailed drawing, the part's material is listed. In general, this does not apply to an assembly drawing; thus, this item may be replaced with the weight of the assembly or some other pertinent information. If you choose to do this, pick Edit Sheet from the pop-down menu under the word Edit at the top of the screen, change the word Material: to Weight (lbs.):, then turn off Edit Sheet by picking it from the pop-down menu again.

Ohio Northern University Mechanical Engr. Dept. Ada, OH 45810	UNLESS OTHERWISE SPECIFIED DIMENSIONS ARE IN INCHES, TOLERANCES ARE .X ±.1 .XX ±.05 .XXX ±.001 ANGLES ±.1°	Material:	Take-up Assembly		
		Drwn By: Your Name goes here			
	DIMENSIONS & TOLERANCES IN ACCORDANCE WITH ASME Y14.2009	Date: 11/25/2023	Scale 1:2	Dwg No. 300000	

The procedure for creating engineering drawings was covered earlier. We will use this procedure to create a drawing template for an assembly or subassembly drawing.

Assembly Drawing Practice

Before creating an assembly drawing, make sure that each item in the assembly has a custom property name called Material. If the item is a part, then the custom property, Material, should be assigned the value of "SW-Material". The Double Quotes are required and the evaluated value will reflect the material assigned to that part. If the item is a subassembly, then the property name, Material, should be assigned the word, Subassembly.

1.  Start SOLIDWORKS if necessary.
2. **File> New…**
3. In Advanced mode, pick A-Size Template, then pick OK.
4. After the drawing border and title block show up, pick the Browse… button in the Model View window.
5. In your working directory, select **"Swinging Link Assembly"**, then pick the Open button.
6. Move the cursor to the lower left area of the border and press the LMB to place the Front View.
7. Move the cursor to the right and press the LMB again to place the Right View.
8. Pick the green checkmark in the Model View window to stop adding views. If the Drawing number shows up in the title block, skip steps 9 through 11.
9. If the Drawing number in the title block is blank, open the Swinging Link Assembly. Under File> pick Properties… In the Properties window, define the Property Name as DrwgNo. Move to the Value/Text Expressions area, then enter the number 411000. Pick the OK.
10. Save the assembly.
11. **File> Close**
12. The drawing number 411000 should now appear in the title block.
13. In the Annotation tab, select the Table icon at the top of the screen, then pick Bill of Materials from the pop-down list.
14. In the Bill of Materials window, the message says, "Select the drawing view to specify the model for creating a Bill of Materials." Pick the Front View, then pick the green checkmark in the Bill of Materials window.
15. Move the Bill of Materials table to the upper right corner of the drawing area, then pick the LMB to place it.

ITEM NO.	PART NUMBER	DESCRIPTION	QTY.
1	Base		1
2	Clevis Pin		1
3	Link		1
4	B27.1 - NA3-50		1

B

16. Pick the "C" column header.

17. 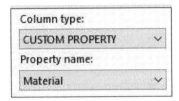 Pick Column Properties from the pop-up menu, then under Property name:, pick Material. The material that each part is made from should appear in the Bill of Materials. Pick anywhere in the graphics area to continue.

| Column type: |
| CUSTOM PROPERTY   ⌄ |
| Property name: |
| Material   ⌄ |

18. Pick Auto Balloons from the top of the screen, then pick the Front view. Balloon numbers 1 through 4 should appear since there are four parts. Pick the green checkmark.

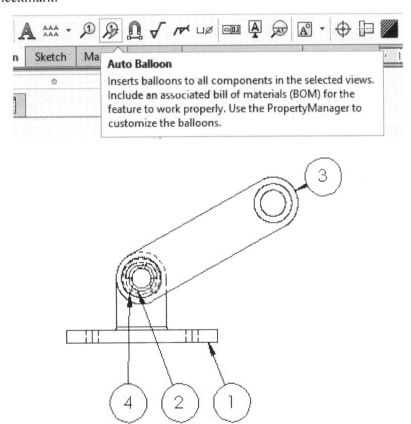

19. Rearrange the balloons so the numbers go counterclockwise around the assembly. Pick a balloon with the LMB, then drag either the balloon or the arrow to a better position.

20. 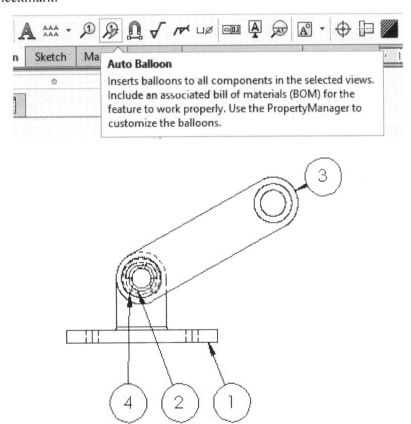 Pick the Center Mark icon from the top of the screen, then pick the hole at the top of the Link and the clevis pin to place center marks. Pick the green checkmark.

21. 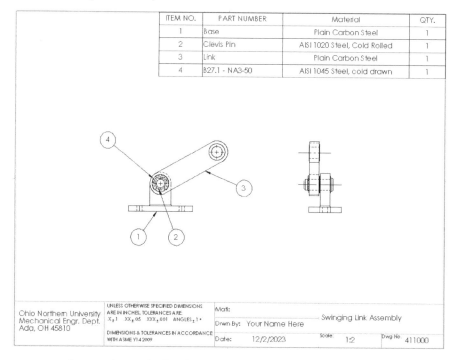 Pick the Centerline icon at the top of the screen, then pick the left side and the right side of the other holes that are viewed from the side in both the Front and Right views. Pick the green checkmark when finished.

22. Save the drawing. Name it **"Swinging Link Assembly Drawing"**.

23. **File> Print...** (if asked to)

ITEM NO.	PART NUMBER	Material	QTY.
1	Base	Plain Carbon Steel	1
2	Clevis Pin	AISI 1020 Steel, Cold Rolled	1
3	Link	Plain Carbon Steel	1
4	B27.1 - NA3-50	AISI 1045 Steel, cold drawn	1

Ohio Northern University Mechanical Engr. Dept. Ada, OH 45810	UNLESS OTHERWISE SPECIFIED DIMENSIONS ARE IN INCHES, TOLERANCES ARE: X±1 XX±.05 XXX±.001 ANGLES±1° DIMENSIONS & TOLERANCES IN ACCORDANCE WITH ASME Y14.2009	Matl:	Swinging Link Assembly
		Drwn By: Your Name Here	Scale: 1:2
		Date: 12/2/2023	Dwg No: 411000

If you want to change the scale for the drawing, do the following.

24. Right-click on Sheet Format 1 in the Feature Manager Design Tree, then pick Properties...

25. In the Scale area of the window, change the Scale to 3:4. Pick the Apply Changes button. The two views are now drawn at ¾ scale.

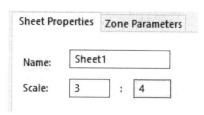

Sheet Properties | Zone Parameters

Name: Sheet1

Scale: 3 : 4

26. Move the views and rearrange the balloons if necessary.
27. Save the drawing. Pick **Save All** if prompted.
28. **File> Print...** (if asked to)

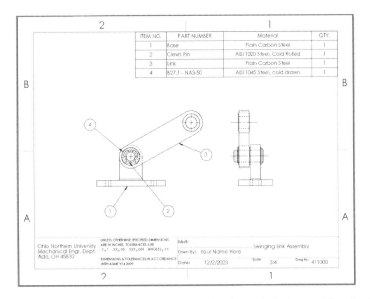

ITEM NO.	PART NUMBER	Material	QTY.
1	Base	Plain Carbon Steel	1
2	Clevis Pin	AISI 1020 Steel, Cold Rolled	1
3	Link	Plain Carbon Steel	1
4	B27.1 - NA3-50	AISI 1045 Steel, cold drawn	1

If you want to create an exploded view of the Swinging Link Assembly, do the following.

29. Pick the Right View, then press <Delete> to remove it from the drawing. When asked to Confirm Delete, pick Yes.

30. Pick the Front View. The Drawing View1 window should appear. Pick the Isometric icon instead of the Front View icon in the Orientation section.

31. In the Reference Configuration area on the left, check the "Show in exploded or model break state" box.

32. Change the Display Style to Shaded if desired or leave it at Hidden Lines Removed.

33. Pick the green checkmark.

34. Move the View so it is in the middle of the drawing area, then rearrange the balloons if necessary.

35. **File> Save As…** Name the drawing, **"Swinging Link Assembly Drawing2"**

36. **File> Print…** (if asked to)

37. **File> Close**

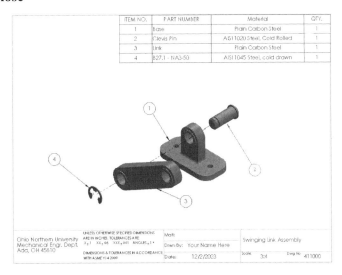

ITEM NO.	PART NUMBER	Material	QTY.
1	Base	Plain Carbon Steel	1
2	Clevis Pin	AISI 1020 Steel, Cold Rolled	1
3	Link	Plain Carbon Steel	1
4	B27.1 - NA3-50	AISI 1045 Steel, cold drawn	1

Assembly Drawing Exercise

Design Intent – Create a 2-view assembly drawing for the Take-up Frame.

1. Start SOLIDWORKS if necessary.
2. **File> New...**
3. In Advanced mode, pick A-Size Template, then pick OK.
4. After the drawing border and title block show up, pick the Browse... button in the Model View window.
5. In your working directory, select **Take-Up Frame**, then pick the Open button.
6. Move the cursor to the lower left area of the border and press the LMB to place the Front view.
7. Move the cursor to the right and press the LMB again to place the Right view.
8. Pick the green checkmark in the Model View window to stop adding views. If the Drawing number shows up in the title block, skip steps 9 through 11.
9. If the Drawing number in the title block is blank, open **Take-Up Frame**. Under File> pick Properties... In the Properties window, define the Property Name as DrwgNo and enter 310001 in the Value/Text Expression area. Pick OK.
10. Save the assembly.
11. **File> Close**
12. The drawing number 310001 should appear in the title block.

In the figure below, the views are not what I want to display.

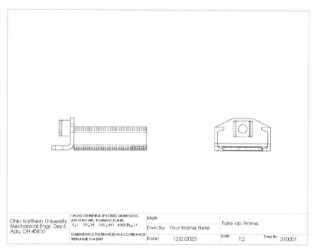

13. Pick the left view. When the Drawing View1 window appears, select the Top view in the Orientation area. When asked, "Do you want to change the drawing view's orientation?" Pick Yes, then pick the green checkmark.

14. 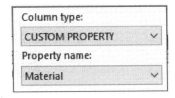 Select the Annotation tab, then select the Table icon at the top of the screen. Pick the Bill of Materials from the pop-down list.

15. In the Bill of Materials window, the message says, "Select the drawing view to specify the model for creating a Bill of Materials." Pick the left view, then pick the green checkmark in the Bill of Materials window.

16. Move the Bill of Materials table to the upper right corner of the drawing area, then pick the LMB to place it.

17. Pick the "C" column header.

18. Pick Column Properties from the pop-up menu, then under Property name: pick Material from the pop-up menu. The material that each part is made from should appear in the Bill of Materials. Pick anywhere in the graphics area to continue.

Column type:

CUSTOM PROPERTY

Property name:

Material

19. Pick Auto Balloons at the top of the screen, then pick the Front view. Balloon numbers 1 through 4 should appear since there are four parts. Pick the green checkmark.

20. Rearrange the balloons by picking a balloon with the LMB, then drag either the balloon or the arrow to a better position so the view looks similar to the figure below.

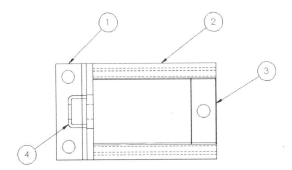

21. Pick the Center Mark icon at the top of the screen, then pick the four circular holes to place a center mark on each. Pick the green checkmark when finished.

22. Pick the Centerline icon at the top of the screen, then pick the two edges of the hole that is viewed from the side in the Lower Plate. Extend the center line so it goes through the Nut Holder hole as well. Zoom In and add centerlines to the 3 holes viewed from the side in the right view. Pick the green checkmark.

23. Save the drawing. Name it **"Take-Up Frame Drawing."**

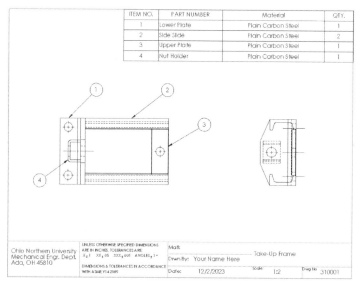

ITEM NO.	PART NUMBER	Material	QTY.
1	Lower Plate	Plain Carbon Steel	1
2	Side Slide	Plain Carbon Steel	2
3	Upper Plate	Plain Carbon Steel	1
4	Nut Holder	Plain Carbon Steel	1

Let's change the scale for the drawing to ¾ scale.

24. Right-click on Sheet Format 1 in the Feature Manager Design Tree, then pick Properties…

25. In the Scale area of the window, change the Scale to 3:4. Pick the Apply Changes button. The two views are now drawn at ¾ scale.

26. Move the views and rearrange the balloons if necessary.

27. Save the drawing. Pick **Save All** if prompted.

28. **File> Print…** (if asked to)

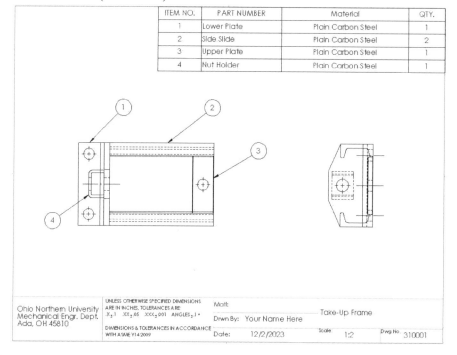

ITEM NO.	PART NUMBER	Material	QTY.
1	Lower Plate	Plain Carbon Steel	1
2	Side Slide	Plain Carbon Steel	2
3	Upper Plate	Plain Carbon Steel	1
4	Nut Holder	Plain Carbon Steel	1

29. **File> Close**

Assembly Drawing Problems

9-1   Create an exploded view subassembly drawing of the **"Take-up Frame"**. Be sure to include the explode lines to show where the parts fit relative to each other. Drawing Number 310001.

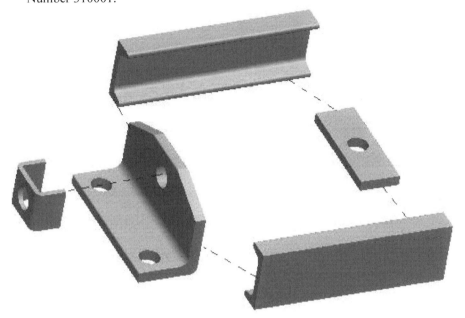

9-2   Create an exploded view subassembly drawing of the **"Take-up Piece"** with Drawing Number 310011. Since it is obvious where the parts go, no explode lines are necessary.

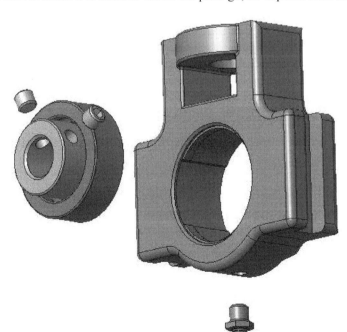

9-3   Create an exploded view assembly drawing of the **"Injector Mold Assembly"**. Be sure to include the explode lines to show where the parts fit relative to each other. Drawing Number 300100.

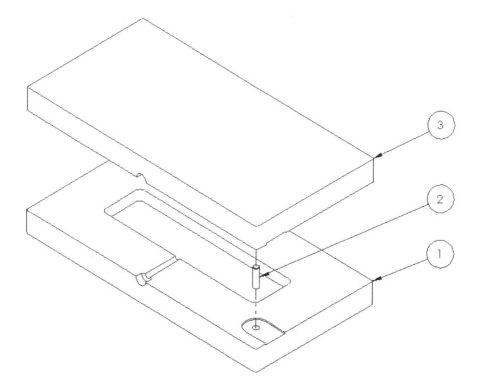

9-4   Create an exploded view subassembly drawing of the **"Step-shaft Assembly"**. Be sure to include the explode lines to show where the parts fit relative to each other. Drawing Number 300204.

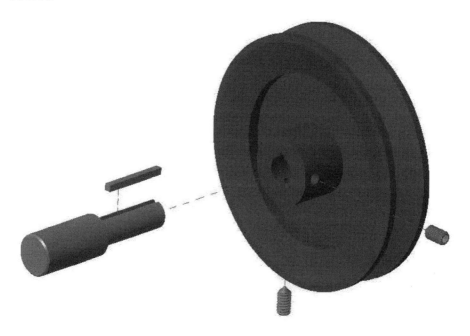

9-5   Create an exploded view assembly drawing of the **"Drill Jig Assembly"**. Be sure to include the explode lines to show where the parts fit relative to each other. Drawing Number 300505.

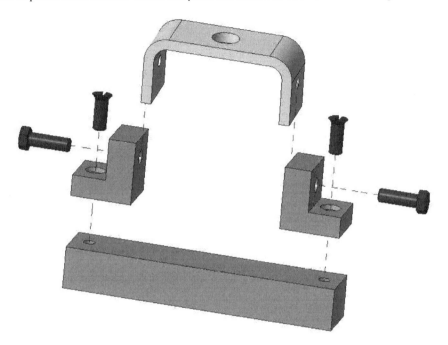

9-6   Create an exploded view assembly drawing of the **"Adjustable Shaft Support"**. Be sure to include the explode lines to show where the parts fit relative to each other. Drawing Number 300606.

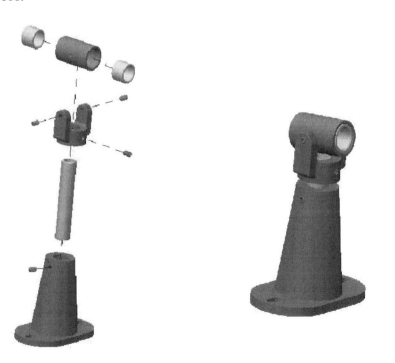

9-7   Create an assembly drawing of the **"Take-up Assembly"**. Drawing Number 300000.

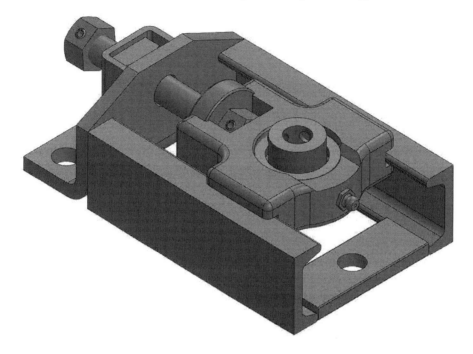

9-8   Create an exploded view assembly drawing of the **"Take-up Assembly"** shown above.
       Drawing Number 300000.

9-9   Create an exploded view assembly drawing of the **"Drilling Jig Assembly"** shown below.
       Drawing number 300909.

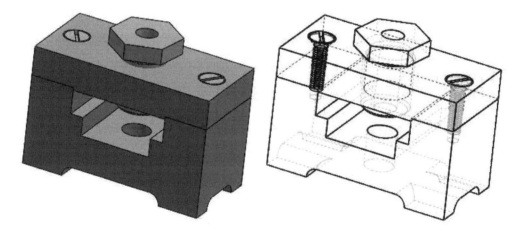

Chapter 10.   Tolerancing and GD&T

Introduction

Tolerance is the amount of variation permitted in the size of a part or the location of a point, axis, or surface. For example, a size dimension given as 1.375 inches ± .004 inches means that the size may be any value between 1.379 inches and 1.371 inches. The tolerance, or total amount of variation, is .008 inches. It becomes the job of the designer to specify the allowable error that may be acceptable for a given feature and still permit the part to function satisfactorily. Since greater accuracy costs more, it is better to specify a generous tolerance.

Without interchangeable part production, the manufacturing industry could not exist, and without effective size and location control by the design engineer, interchangeable manufacturing would not happen. For example, a car manufacturer subcontracts the production of many of the car's parts to other companies. These parts plus replacement parts must be similar so that any one of them will fit properly in the car. It might be thought that if the dimensions are given on the blueprint, such as 4½, 3.125, etc., all the parts will be exactly alike and will fit properly. Unfortunately, it is impossible to make any part to an exact size or locate a feature of size exactly. The part can be made very close to the specified dimensions, but such accuracy is expensive. Luckily, exact sizes are not required, only varying degrees of accuracy according to the functional requirements. The maker of children's building blocks would soon go out of business if they insisted on making the blocks with high accuracy. No one would be willing to pay such a high price for building blocks. What is needed is a means of specifying dimensions with whatever accuracy is required. The answer to this situation is the specification of some tolerance on each dimensional value.

To control the dimensions of two parts so that any two mating parts will be interchangeable, it is necessary to assign tolerances to the dimensions, as shown below. The diameter of the hole must be machined to a diameter between 1.250 inches and 1.252 inches. These two values become the acceptable dimension limits and the difference between them, .002 inches, becomes the allowance. The shaft must be produced between the limits of 1.248 inches and 1.246 inches in diameter. The shaft allowance is the difference between these two values or .002 inches.

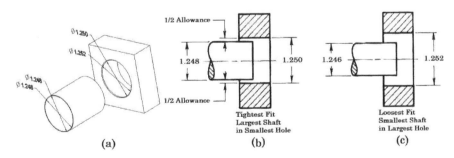

The maximum and minimum shaft diameters are listed on the left. The loosest fit or maximum clearance occurs when the smallest shaft is mated with the largest hole. The tightest fit or minimum clearance occurs when the largest shaft is mated with the smallest hole. The allowance is the difference between these two dimensions on the tightest fit or .002 inches. Any shaft will fit any hole interchangeably.

Before continuing, let's define some terms.

> Nominal Size—the value that is used for the general identification of a size.

> Basic Size—the value from which limits of size are derived by applying the allowance and tolerances. It is the theoretical value from which the size limits are calculated.

> Limits—the maximum and minimum values indicated by a tolerance dimension.

> Allowance—the minimum clearance or maximum interference between two mating parts. The allowance represents the tightest permissible fit and is simply the smallest hole minus the largest shaft. For clearance fits, this difference will be positive. For interference fits, the allowance will be negative.

There are three general types of fits between parts:

1. Clearance fit—an internal member fits into an external member leaving an air space or clearance between the parts. The largest shaft is 1.248 inches and the smallest hole is 1.250 inches, which gives the smallest acceptable air space between the parts as .002 inches. This space is the allowance. In a clearance fit, the allowance is positive.

2. Interference fit—the internal member is larger than the external member such that there is always interference. If the smallest shaft is 1.506 inches, and the largest hole is 1.502 inches, there would be an interference of .004 inches.

3. Transition fit—the fit might be either a clearance fit or an interference fit. If the smallest shaft was 1.503 inches and the largest hole was 1.506 inches, there would be .003 inches to spare (positive allowance). However, if the largest shaft, 1.509 inches, was forced into the smallest hole, 1.500 inches, an interference (negative allowance) of .009 inches would occur.

If allowances and tolerances are properly specified and features of size are located properly, mating parts will be interchangeable. However, for close fits, it may be necessary to specify very small tolerances and allowances, thus the cost increases. To avoid this expense, selective assembly is used. In selective assembly, all parts are inspected and classified into several classes according to their actual sizes, so that "small" shafts can be mated with "small" holes, "medium" shafts with "medium" holes, and "large" shafts with "large" holes. In this way, satisfactory fits may be obtained with less expense.

Basic Hole System

Standard broaches, reamers, and other standard tools are used to make holes. Standard plug gages are used to check their size. Shafts can be machined to any desired diameter. Therefore, toleranced dimensions are commonly calculated based on the basic hole system. In this system, the minimum hole is taken as the basic size, and a plus or minus allowance is assigned to the shaft. Then the hole and shaft tolerances are applied to these parts.

The minimum size of the hole shown below is .500 inches and is taken as the basic size. An allowance of .003 inches is decided upon and subtracted from the basic hole size, giving the maximum shaft diameter of .497 inches. Tolerances of .003 inches for the hole and .002 inches for the shaft are applied to obtain the maximum hole of .503 inches and the minimum shaft of .495 inches. Thus, the minimum clearance (allowance) between the parts becomes .500 inches − .497

inches = .003 inches (smallest hole minus largest shaft), and the maximum clearance is .503 inches – .495 inches = .008 inches (largest hole minus smallest shaft).

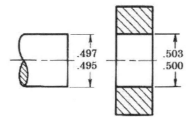

For an interference fit, the maximum shaft size is calculated by adding the desired maximum interference to the basic size. The basic size is 1.2500 inches. The maximum interference was set to .0024 inches, which is added to the basic size making the largest shaft 1.2524 inches.

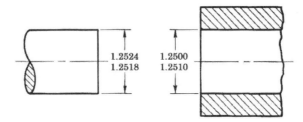

For common-fractional dimensions, the operator is not expected to work closer than he can be expected to measure with a steel ruler. It is customary to indicate an overall general tolerance for all common-fraction size dimensions using a note in the title block, such as ALL FRACTIONAL DIMENSIONS ±1/64″ UNLESS OTHERWISE SPECIFIED. General angular tolerances may be given as ANGULAR TOLERANCE ±1°. Without GD&T control feature frames, tolerances on decimal size dimensions may be given in a general note, such as:

SIZE TOLERANCES ARE:
.X ± .1
.XX ± .005
.XXX ± .001

Every dimension on an engineering drawing must have a tolerance, either direct or implied by a general tolerance note. Two methods of expressing tolerances on size or location dimensions follow:

1. Limit Dimensioning—the preferred method, specifies the maximum and minimum limits of size and location. Generally, the maximum material condition limit is placed above the minimum material condition limit. In note form, the maximum material condition limit is given first, thus for a hole: Ø.500- Ø.502. For a shaft, the note would read as Ø.504- Ø.503. (Note that many CAD systems always show the largest value first.)

2. Plus and Minus Dimensioning—the basic size is followed by a plus and minus expression of the tolerance resulting in either a unilateral (one nonzero value) or bilateral (two nonzero values) tolerance. If two unequal tolerance numbers are given, one plus and one minus, the plus value is placed above the minus. One of the tolerances may be zero. If a single tolerance value is given, it is preceded by the plus-or-minus symbol. This method can be used when the plus and minus values are equal. The unilateral system of tolerances allows

variations in only one direction from the specified size. This method is advantageous when a critical size is approached as the material is removed during manufacturing, as in the case of close-fitting holes and shafts. The bilateral system of tolerances allows variations in both directions from the specified size. *This method was typically used before GD&T to locate features of size; however, it does not account for orientation or form errors.*

When using either method above, it is important to be aware of the effect of one tolerance on another. When the location of a surface is affected by more than one tolerance, the tolerances are accumulative. In the figure below, if dimension C is omitted, surface Y would be controlled by both dimensions A and B, and there could be a total variation of .010 inches instead of the variation of .005 inches permitted by dimension C. If the part is made using the minimum tolerances of A, B, and C, the total variation in the length of the part from 3.750 inches would be .015 inches; the part could be as short as 3.735 inches. However, the tolerance on the overall dimension D is only .005 inches, thus allowing the part to be only as short as 3.745 inches. Note that the part is over-dimensioned as shown. In some cases, for functional reasons, it may be desired to hold all three small dimensions A, B, and C close without regard to the overall length. In such a case, the overall dimension should be marked as a reference (REF). It may be desirable to hold the two small dimensions, A and B, and the overall dimension close without regard to dimension C. In this case, dimension C should be omitted or marked as reference (REF).

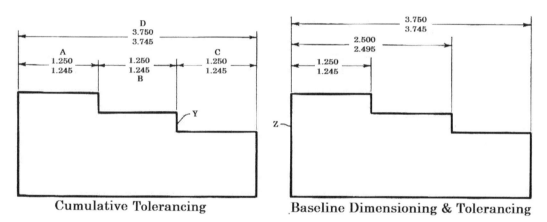

Cumulative Tolerancing   Baseline Dimensioning & Tolerancing

It is better to dimension each surface so that it is affected by only one dimension and tolerance. This is done by referencing all dimensions from a single datum surface, like Z, shown above.

As has been stated previously, size tolerances should be as large as possible and still permit the proper functioning of the part. This allows for less expensive tools, lower labor and inspection costs, and reduced material scrap, thus reducing the overall cost of the part. A chart of some shop processes with expected tolerances is shown next.

Shop Process	Tolerance Range for Various Shop Processes								
Drilling									
Milling									
Turning, Boring									
Planing, Shaping									
Reaming									
Broaching									
Grinding, Boring									
Lapping, Honing									

From	Up to	Tolerances (inches)								
0.000	0.600	0.00015	0.0002	0.0003	0.0005	0.0008	0.0012	0.002	0.003	0.005
0.600	1.000	0.00015	0.00025	0.0004	0.0006	0.0010	0.0015	0.0025	0.004	0.006
1.000	1.500	0.0002	0.0003	0.0005	0.0008	0.0012	0.002	0.003	0.005	0.008
1.500	2.800	0.00025	0.0004	0.0006	0.0010	0.0015	0.0025	0.004	0.006	0.010
2.800	4.500	0.0003	0.0005	0.0008	0.0012	0.002	0.003	0.005	0.008	0.012
4.500	7.800	0.0004	0.0006	0.0010	0.0015	0.0025	0.004	0.006	0.010	0.015
7.800	13.600	0.0005	0.0008	0.0012	0.002	0.003	0.005	0.008	0.012	0.020
13.600	21.000	0.0006	0.0010	0.0015	0.0025	0.004	0.006	0.010	0.015	0.025

Coordinate Tolerancing Issues (Use GD&T instead)

Coordinate tolerancing can be applied to the location of a hole from a specified reference plane. It is important to determine the location of the reference planes or datum planes before dimensioning the part. Reference planes must be machined surfaces, the centerline of a machined hole, or the center plane of a feature of size. In the figure below the left and bottom edges of the part are defined as two of the part's datum planes. The center of the hole is located relative to these two datum planes. The center of the hole can be anywhere inside the square region of .010 inches by .010 inches. The positional tolerance of the hole is the length of the diagonal of this square region or .014 inches. If the maximum size of the bolt to go into this hole and a corresponding hole in a similar part is .500 inches, the hole diameter should be at least .514 inches to avoid interference. In equation form, H = S + T, where:

T = maximum positional tolerance of the hole's center = sqrt $(.010^2 + .010^2)$ = .014
S = maximum diameter of the bolt or shaft = .500
H = minimum diameter of the hole so there is no interference
H = .500 + .014 = .514 inches

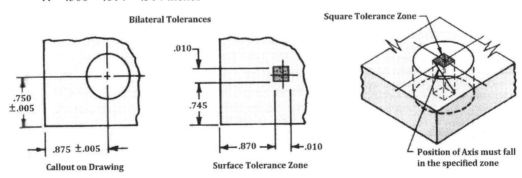

The same calculations would hold true if you were locating a second clearance hole relative to the first, as shown next. The above equation can also be used to determine the allowable tolerance on the position of the second hole's center if the maximum diameter of the bolt and the minimum diameter of the hole are known, the holes are perfectly round, the holes are perpendicular to the surface, and the mating surfaces are perfectly flat. (*WOW, I see an issue with this last statement.*)

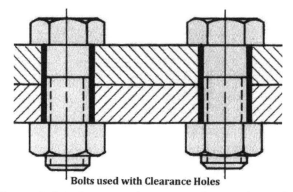

Bolts used with Clearance Holes

When one of the mating parts is to be screwed to a second part using a threaded hole and a cap screw or stud, the method is called fixed-fastener. When the threaded fastener is the same basic size as the clearance hole the positional tolerance for the two holes' centers becomes:

$$T = (H - S)/2.$$

The hole's minimum diameter must be at least: $H = S + T_H + T_S$

where:

H = minimum diameter of hole so there is no interference if perfectly formed
S = maximum diameter of bolt or shaft
T_H = clearance hole's positional center tolerance
T_S = shaft's positional center tolerance (threaded hole's center)

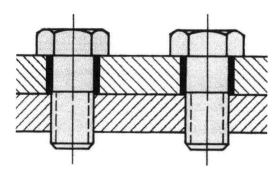

A lid is to be screwed onto a box using ½-13UNC-2A × 1.00 long cap screws as shown above. The distance between the two threaded holes is $2.000 \pm .003$ inches. The clearance holes are located $2.000 \pm .004$ inches apart. The maximum size for the cap screw is .500 inches. Determine the minimum diameter for the clearance holes so that there is no interference if the holes are perfectly round and perpendicular to the mating surfaces. (*Note that we are ignoring the misalignment of the holes into or out of the page so this may not assemble.*)

$H = S + T_H + T_S$
$H = .500 + .008 + .006$
$H = .514$ inches in diameter (minimum)

Geometric Tolerances

This chapter is not meant to be a comprehensive coverage of Geometric Dimensioning and Tolerancing (GD&T), but rather a simple introduction to it and how to add GD&T to your engineering drawings as needed. You might ask, "Why GD&T?" Geometric Dimensioning and Tolerancing are recognized around the world as the only effective way to define a part geometry.

Geometric tolerances are used to control location, orientation, and form. "Tolerance of form" specifies how far surfaces are permitted to vary from the perfect geometry implied by the engineering drawings. Theoretically, planes, cylinders, cones, etc. are perfect forms, but since it is impossible to produce a perfect form, it is necessary to specify the amount of variation permitted. Geometric tolerances define such conditions as flatness, parallelism, perpendicularity, angularity, cylindricity, and straightness. When geometric tolerances are not indicated on the drawing, the actual part is understood to be acceptable if it is within the dimensional limits shown, regardless of variations in form.

In the case of fabricated bars, sheets, and tubing established industry standards prescribe acceptable conditions of straightness, flatness, etc., and these standards are understood to hold if geometric tolerances are not shown on the drawing.

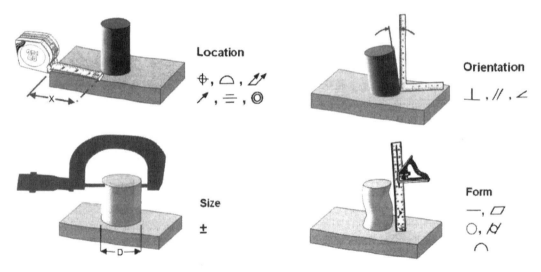

Let's look at a simple example using a positional GD&T. A steel block is to be bolted into the corner of a steel box using a ½-13UNC-2A bolt. The clearance hole for the bolt has a minimum diameter of ½ inch +1/64-inch or .516 inches. Determine the positional tolerances for the clearance hole.

Rearranging the positional tolerance equation and solving for the positional tolerance becomes:

$$T = H - S$$
$$T = .516 - .500$$
$$T = .016 \text{ inches (diameter)}$$

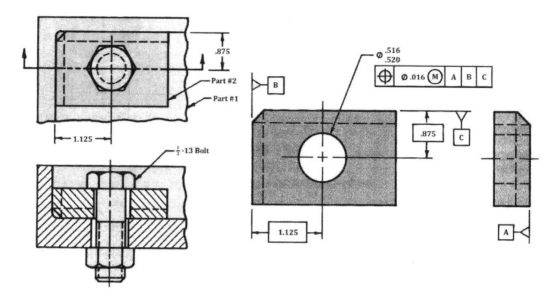

Basic dimensions represent exact theoretical values. This is the basis from which permissible variations are allowed. They are enclosed in a rectangular box and shown without tolerances. The corresponding control frame shows the allowable tolerance. Since we are locating the hole using GD&T, the .875-inch and 1.125-inch dimensions shown on the left above will become basic dimensions in the figure on the right above. Also, these basic dimensions need to be given relative to Datums A, B, and C. The left and top edges of the block are shown as Datum planes B and C. Datum plane A is located where the block would sit on the box and is specified as the primary datum.

The hole is dimensioned using limit dimensions with the maximum material condition (MMC) dimension given first. The hole can be as small as .516 inches but must be less than .520 inches. (Many CAD systems list the larger value first.) The control block under this dimension is the GD&T. The first area of this control block represents the type of geometric tolerance; in this case, it is a positional tolerance. The second area gives the allowable tolerance for the position with an additional qualifier. The center of the hole must be located within a .016-inch diameter of its exact center location at maximum material condition. The next three areas represent the order in which the block is placed on the datum surfaces during its quality inspection. The block is set on Datum A, then pushed left toward Datum B, and finally pushed back toward Datum C.

Feature-control symbols should be used in place of notes to specify positional and form tolerances. The symbols for the geometric characteristic in which the tolerances apply are shown on the following page.

FEATURE	TOLERANCE TYPE	CHARACTERISTIC	SYMBOL
Individual Features	Form	Straightness	⎯
		Flatness	▱
		Circularity (Roundness)	◯
		Cylindricity	⌭
Individual or Related Features	Profile	Profile of a Line	⌒
		Profile of a Surface	⌓
Related Features	Orientation	Angularity	∠
		Perpendicularity	⊥
		Parallelism	∥
	Location	Position	⊕
		Concentricity	◎
		Symmetry	⌯
	Runout	Circular Runout	*↗
		Total Runout	*↗↗

Before we begin an exercise, we need to review a few features of geometric tolerancing. Straightness controls how close to a straight line an edge or surface must be to be acceptable. For a cylinder, it controls the straightness of its axis center, not the straightness of the surface. The straightness tolerance must be less than the size tolerance.

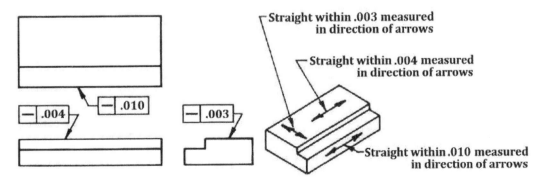

Flatness controls how close to a flat plane the surface must be to be acceptable. The flatness tolerance must be less than the size tolerance. Flatness per unit area can also be specified when it is necessary to control the flatness in a given area closer than the overall flatness requirement.

See Figure 12-14. The overall flatness of the surface must be within .009 inches; however, the flatness of any given square inch must be within .002 inches.

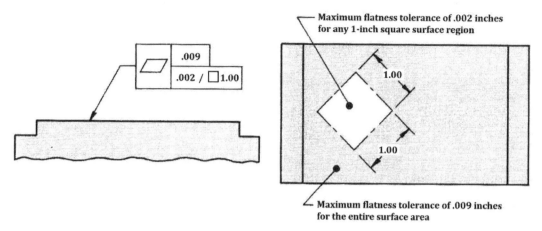

"Features of size" reference shafts, holes, and parallel surfaces. With features of size, the control frame is associated with the size dimension. For example, if we wanted to control the straightness of a shaft's axis center, we would use the straightness symbol as before and include a diameter symbol with the specified form tolerance. We may also add a modifier such as MMC (maximum material condition), LMC (least material condition), or RFS (regardless of feature size). RFS is the default modifier if none is shown.

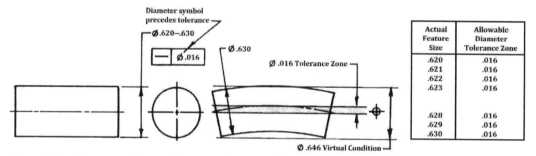

Geometric tolerances for straightness, flatness, circularity, and cylindricity do not need to be located relative to any datum planes; however, geometric tolerances for orientation and position need to be referenced from specified datum planes. These datum planes must be specified on the drawing. The primary datum is a flat surface against which the part is first placed. The part is moved so that it touches the secondary datum next. Finally, the part is moved perpendicular to the previous motion until it contacts the tertiary datum. Note that a hole may be used as a datum. If a single datum is established using two other datums, then both datums are referenced in the same area of the control frame with a dash between them.

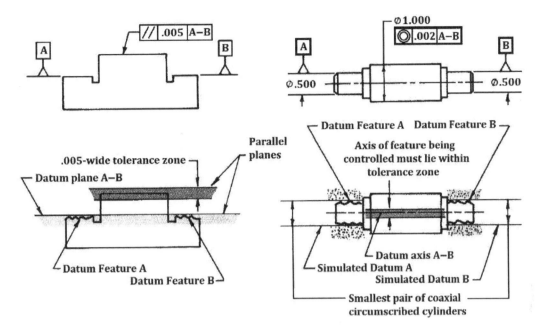

Orientation refers to the angular relationship between two lines or surfaces. Orientation tolerances control angularity, perpendicularity, and parallelism. A tolerance of form or orientation should be specified when the size and location tolerances do not provide proper control of the feature. **A perpendicularity tolerance should be used on the secondary and tertiary datum planes to indicate how close to perpendicular they need to be relative to Datum A.**

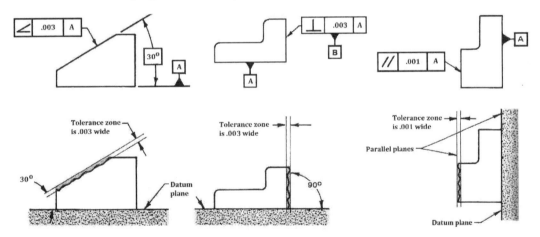

Positional tolerances are used on groups or patterns of holes at MMC. This method meets the functional requirements for the feature in most cases. Positional tolerancing refers to the tolerance zone away from the exact location of the centerline for the hole or shaft. At MMC the tolerance cannot exceed the value specified; however, the positional tolerance allowed is dependent upon the actual size of the considered feature. If the holes are made at MMC, the smallest acceptable (.625 inches), then the hole centers must be dead on location. If a hole is .005 inches larger at .630 inches, then the center location for this hole may be off by .0025 inches or half of the hole size tolerance. (See the figure on the following page.)

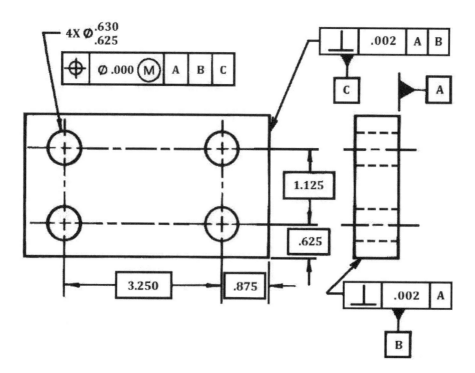

Circularity tolerance is measured radial and specifies the distance between two circles that a particular circular cross-section must stay in for it to be acceptable. It is shown without modifiers; thus it is assumed to be regardless of feature size (RFS).

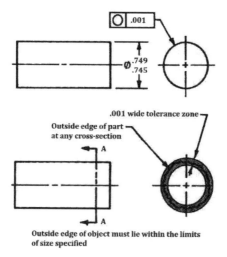

That's enough review for now. For a more detailed look at GD&T see "Geometric Dimensioning and Tolerancing," based on ANSI/ASME Y14.5M-1994, by David Madsen, or "The GD&T Hierarchy, Y14.5M-2009," by Don Day. Let's get started.

We will create the part, and then make an engineering drawing for the part with some appropriate geometric tolerances.

Tolerancing and GD&T Practice

Before beginning this section, be sure "base.SLDPRT" from the Chapter 8 assembly practice session is in your working directory. Note that a third view is required to show the rounded corners.

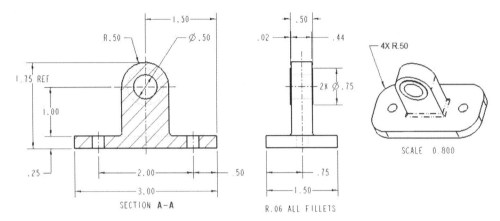

Design Intent – Create an engineering drawing of "Base.SLDPRT" with its border and title block. Use GD&T to make sure the ½-inch hole is positioned within a diameter of .003 inches relative to Datums A, B, and C. The bottom of the Base needs to be flat within .005 inches. Also, the first ¼-inch hole needs to be perpendicular to the bottom of the base within a diameter of .004 inches, the second ¼-inch hole needs to be positioned within a .004-inch diameter relative to the bottom of the base, and the first ¼-inch hole when it is at its smallest size or at its Maximum Material Condition (MMC).

1. Start SOLIDWORKS if necessary.
2. Change to Advanced mode if not already there.
3. Pick the SolidWorks Templates tab.
4. Pick the A-Size Template under the SolidWorks Template tab. Pick **OK**.
5. Browse through the SolidWorks Parts folder until you locate **Base**, then pick Open.
6. Move the cursor to locate the FRONT view, then press the LMB.
7. Move the cursor to the right to place the RIGHT side view, then press the LMB again. Move above the Front View and press the LMB again to add the TOP view. Press <Esc> to terminate the view placement procedure.

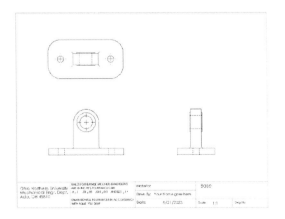

8. The drawing number and the Material section of the title block are probably blank so we need to add these two custom property names to Base.SLDPRT. Open Base.SLDPRT and add the Property Name Material as "SW-Material" along with DrwgNo as 411002, then save the part and close it.

	Property Name	Type	Value / Text Expression	Evaluated Value
1	Material	Text	"SW-Material"	Plain Carbon Steel
2	DrwgNo	Text	411002	411002

9. When you return to the drawing, the drawing number and the material the base is made from should show up in the title block.

10. Pick Drawing View1 from the Feature Manager Design Tree, then press and hold the RMB until a pop-up menu appears. Pick **Tangent Edges Removed** under the Tangent Edge option to remove tangent lines that are in line with the hole's centerlines.

11. Do the same thing for Drawing View2.

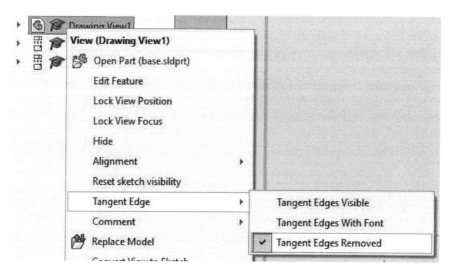

12. ⊡ With the Annotation tab selected, pick the Centerline tool. Add centerlines to the rectangular view of each of the holes. Pick the green checkmark when finished.

13. Extend the centerlines for the round holes similar to the figure below.

14. ⟋ With the Annotation tab selected, pick the Model Items icon.

15. For the Source: pick Entire Model, then pick the green checkmark.

16. Rearrange the dimensions that showed up by dragging some of the dimensions off of the part.

17. Based upon the Contour principle, move some dimensions to a different view that better reflects the feature.

18. Dimension the three views according to the standards listed in Chapter 6.

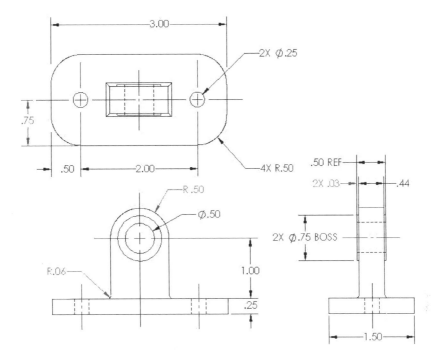

19. Since we want to get rid of the R.06 dimension on the Front view, move the cursor over it without clicking. The variable name for the dimension will appear. In my case, it was "RD2@Drawing View1 of Base".

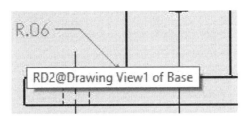

20. **A** Pick the Note icon at the top of the screen, move to the lower right of the drawing area, then press the LMB. Type: R"RD2@Drawing View1" ALL FILLETS. The double quotes are required. Pick the green checkmark to accept the Note. Delete the R.06 dimension from the Front View since we have replaced it with a general note above the title block.

21. Save the drawing as "**Base Drawing**".

22. Select the GD&T icon, then pick the bottom surface of the Front view, then press the LMB to place it. Pick the Flatness icon. Set the Range to .005, then pick Done. The bottom surface must be flat within .005 inches.

23. Pick the Datum icon, then pick the Flatness GD&T box you just created. The letter A should appear in the Datum box. Move below the GD&T box and press the LMB to place the Datum A marker. Press <Esc>.

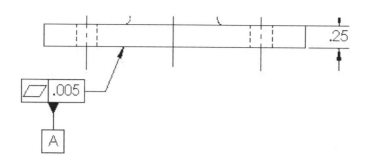

24. Pick the 2X ϕ.25 dimension in the Top view to bring up the Dimension window. In the Tolerance/Precision area, select Limit from the pull-down menu. Enter 0.003 for the plus value. Leave the minus value at zero. Set the precision to three decimal places. Pick the green checkmark.

25. ▣ Select the GD&T icon, then pick the .253/.250 dimension.

26. ⊥ Pick the perpendicular icon for the GD&T symbol. In the Tolerance window, pick the diameter symbol, then enter a value of .002. Pick the Add Datum button to add Datum A. Pick Done, then pick the green checkmark. The center of this ¼-inch hole has to be perpendicular to the bottom surface of the Base within a diameter of .002 inches.

27. 🄰 Pick the Datum icon, then pick the perpendicular GD&T box you just created. Move below the GD&T box and press the LMB to place the Datum B marker. Press <Esc>. Pick the green checkmark to complete this Datum Feature

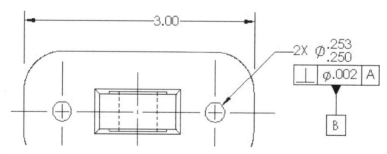

28. 💾 Save the drawing. Pick the Save All button to save the part as well.

29. ▣ Select the GD&T icon, then pick the other ¼ inch hole.

30. ⊕ Pick the position icon for the GD&T symbol. In the Tolerance window, pick the diameter symbol (ϕ), then enter a value of .004.

31. Pick the Add Datum button to add Datum A.

32. Ⓜ Pick the Add New button, to add Datum B at Maximum Material Condition. Pick Done. Pick the green checkmark, then reposition the GD&T box.

33. Pick the Datum icon, then pick the GD&T box you just created. The letter C should appear in the Datum box. Move below the GD&T box, then press the LMB to place Datum C. Pick the green checkmark.

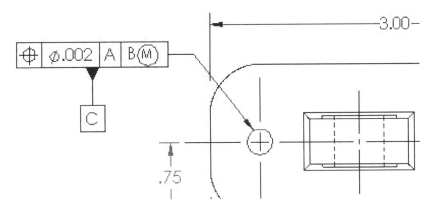

34. Pick the 2.00 dimension between the two holes. In the Tolerance/Precision area, select Basic from the pull-down menu. A box should appear around the 2.00 dimension. The theoretical position of the center of the second ¼-inch hole is exactly 2.00 inches from the first ¼-inch hole. A diameter tolerance of .002 inches is allowed for the misplacement of its center when the first ¼-inch hole is at its smallest value or MMC.

35. Pick the ϕ.50 dimension in the Front view to bring up the Dimension window. In the Tolerance/Precision area, select Limit from the pull-down menu. Enter 0.006 for the plus value. Leave the minus value at zero. Set the precision to three decimal places. Pick the green checkmark.

36. Select the GD&T icon, then pick the .506/.500 dimension.

37. Pick the position icon for the GD&T symbol. In the Tolerance window, pick the diameter symbol (ϕ), then enter a value of .003.

38. Pick the Add Datum button to add Datum A.

39. Pick the Add New button, then enter B to add Datum B. Pick the MMC icon for Maximum Material Condition.

40. Pick the Add New button, then enter C to add Datum C. Pick Done, then pick the green checkmark. Reposition the dimension along with its GD&T for its position.

41. Save the drawing. Pick the **Save All** button to save the modified part as well if asked.

When you look at the top .50-inch diameter hole, there is no distance between Datum B and the hole. We are missing a needed dimension.

42. Pick the Horizontal Dimension icon under the Smart Dimension icon.

43. Pick the right ¼-inch hole's centerline and the center hole's centerline in the Top view. Move down and place the dimension using the LMB. Press <Esc>.

44. Pick the 1.00 dimension just created, then in the Tolerance/Precision area, pick Basic from the pull-down menu. Pick the green checkmark.

45. Move the Top view up a bit or the Front view down a bit if necessary so the dimensions are not crowded. Your drawing should look similar to the figure below.

Because the part was created about the Mid Plane, there is still another dimension missing.

46. Use the Smart Dimension tool to add a .53 dimension between the back of the Base and the edge of the center extrusion in the Right view.

47. Save the drawing.

48. **File> Print...** (if asked to)

49. **File> Close**

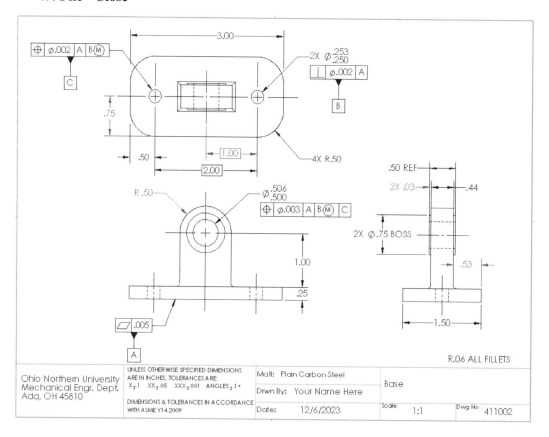

Design Intent - Create a 12-inch by 9-inch by 1.00-inch cover plate made of AISI 1020 steel. Name the part Cover Plate-9x12. Add a 15/16-inch diameter clearance hole (limits of .941 and .937) in each of the four corners 2.00 inches from each edge. Place 2-inch rounds in each of its four corners. Round the upper lip of the cover plate using a 0.12-inch radius. The four 7/8-inch studs that go through these four clearance holes have a maximum diameter of .875 inches and the same positional tolerance as the clearance holes. Add GD&T to your drawing. The bottom of the Cover Plate needs to be flat within .001 inches. The left edge of the Cover plate needs to be perpendicular to the bottom within .003 inches. The back edge of the Cover Plate needs to be perpendicular to the bottom and the left side of the Cover Plate within .003 inches. Drawing number 212120.

1. Start SOLIDWORKS if necessary.
2. **File> New…** Pick **Part_IPS_ANSI** under the SolidWorks Templates. Pick OK.
3. Sketch the 9-inch x 12-inch Cover Plate on the Front plane.
4. Extrude it to 1 inch.
5. Add 2.00-inch rounds in each of the four corners.
6. Add a 15/16-inch hole 2.00 inches from the left and top edge.
7. Use the Pattern tool to create the 4 needed holes.
8. Add an R.12-inch fillet around the top edge of the part.
9. Save the part as **"Cover Plate-9x12"**.
10. **File> Close**

11. **File> New…** Pick the **A-size Template** under the SolidWorks Templates. Pick OK.
12. Pick the Browse… button, then locate **Cover Plate-9x12** in your working directory.
13. Use the LMB to place the Front view and the RIGHT side view, then pick the green checkmark.
14. Right-click on the Front view, then pick Tangent Edge and Tangent Edge Removed. Do the same for the Right view. Then pick the green checkmark.
15. Extend the centerlines for the four holes.
16. Model Items Dimension tool, then dimension the two views similar to the following figure. Use the Dimension tool to add the 2.00-inch dimensions in the upper left corner of the Front view.

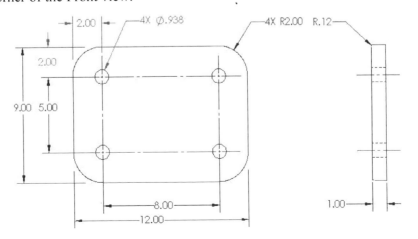

17. Add the three necessary Datum A, B, and C. Be sure Datum B and Datum C do not align with any dimension lines because if you align the datum with a dimension line, the datum refers to the center axis of the dimensioned feature.

18. Pick 2.00, 5.00, and 8.00 dimensions and make them Basic dimensions.

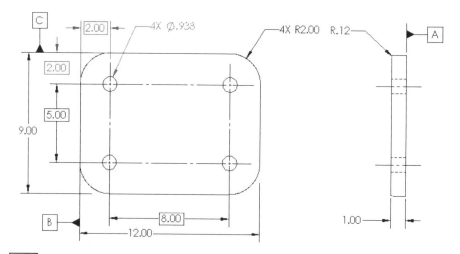

19. Use the GD&T tool to add a flatness tolerance of .001 inches to the bottom surface of the part. We could have created this GD&T box first, then attached Datum A to the GD&T box.

20. Use the GD&T tool to add a perpendicular tolerance of .003 inches to the left edge of the part relative to Datum A.

21. Use the GD&T tool to add a perpendicular tolerance of .003 inches to the top edge of the part relative to Datum A first, then Datum B.

22. Pick the 4X hole dimension, then set its Limits to .941 and .937 inches.

23. How big of positional tolerance can we use and still slide the Cover Plate-9x12 over the four threaded studs? Let's calculate it now. Let's assume the positional tolerance of the holes is the same as for the threaded studs. (See the previous Tolerancing and GD&T Explored section.)

H = minimum diameter of the hole so there is no interference if perfectly formed
S = maximum diameter of the stud = 7/8 inch = .875 inch
T_H = clearance hole's positional tolerance
T_S = stud's positional tolerance (assume same as the hole's clearance)
H = S + T_H + T_S = .875 + T_H + T_S = .937 inches
T_H = T_S = .031 inches (maximum for assembly)

24. Use the GD&T tool to add a positional tolerance of .030 inches to the four holes. Since the part is placed on the stud surface, then pushed back against the top edge of the part, and finally slid left to the left edge of the part, the Datums need to be listed as A, C, and then B.

25. Save the drawing. Name it **"Cover Plate-9x12 Drawing"**. Since the tolerances affected the part, pick the Save All button as well to save the part.

26. Note that the material and the drawing number are missing from the title block.
27. **File> Open…** Pick the Cover Plate-9x12 from the list of parts.
28. Right-click on Material <not specified>, and pick Edit Material. Pick AISI 1020 steel from the list. Pick the Apply button. Pick the Close button.
29. **File> Properties**
30. For the Property Name, enter Material, then enter "SW-Material" in the Value/Text Expression area. Double quotes are required.
31. For the Property Name, enter DrwgNo, then enter 212120 in the Value/Text Expression area. Pick OK.
32. **File> Close**
33. Go back to the drawing and verify that the material is shown as AISI 1020 in the material area of the title block and that the drawing number is 212120.
34. **File> Print…** (if asked to)
35. File> Close

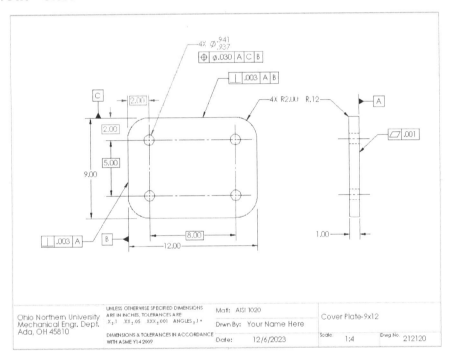

Tolerancing and GD&T Exercise

Design Intent – Create an 8.00-inch diameter .75-inch thick round cover with six equally spaced ½-inch diameter holes on a bolt circle of 6.00 inches. The cover plate is made from plain carbon steel. The six holes, having an upper limit of 0.536 inches and a lower limit of 0.531 inches and must be countersunk at 82 degrees to accommodate ½-13UNC-2A flathead screws. The cover plate's outer diameter must have an upper limit of 8.000 inches and a lower limit of 7.995 inches. Round the top edge of the cover plate using a .06-inch radius. Datum A is the bottom surface of the cover plate and must be flat within 0.001 inches. Datum B is the outside diameter of the cover plate and must be within 0.002 inches of perpendicular relative to the bottom surface, Datum A. Datum C is one of the holes in the pattern. The hole pattern must be located within a diameter of 0.002 inches at MMC relative to Datums A and B. Also, the holes need to be perpendicular to Datum A within .001 inches. Show the cover plate round in the Front view on the drawing. Assign a drawing number of 200120.

1. Start SOLIDWORKS if necessary.
2. **File> New…** Pick **Part_IPS_ANSI** under the SolidWorks Templates. Pick OK.
3. Sketch the Round Cover Plate on the Front plane.
4. Extrude it to a depth of .75 inches.
5. Add an R.06-inch fillet around the top edge of the part.
6. Create a countersunk hole for a ½-inch flat screw using the bolt pattern of 6.00 inches. The countersink needs to be at 82 degrees.
7. Use the Radial Pattern tool to create the 6 equally spaced holes.
8. **File> Properties**
9. Assign Plain Carbon Steel as the material, and assign a drawing number of 200120 as the part properties. Pick OK.

	Property Name	Type	Value / Text Expression	Evaluated Value
1	Material	Text	"SW-Material"	Plain Carbon Steel
2	DrwgNo	Text	200120	200120

10. Save the part as **"Round Cover Plate"**.

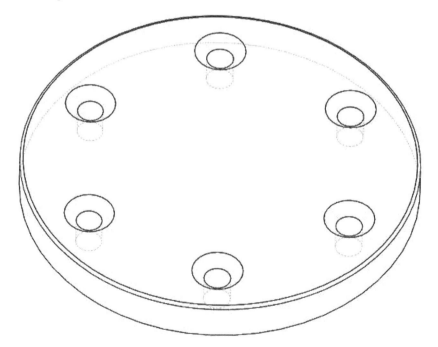

11. **File> New…** Pick the **A-size Template** under the SolidWorks Templates. Pick OK.
12. Double-click on the **Round Cover Plate** in *Part/Assembly to Insert* area.

13. Use the LMB to place the Front view and the RIGHT side view, then pick the green checkmark.

14. ⊞ Add centerlines to the four holes in the Right view. Also, add a centerline to the 8.00-inch diameter.

15. Dimension the two views similar to the following figure.

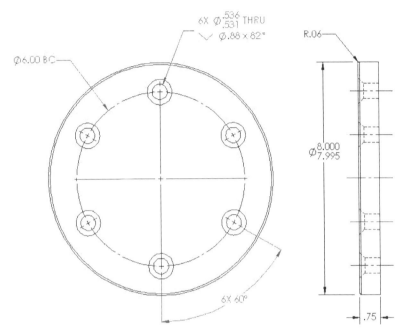

16. Save the drawing as **"Round Cover Plate Drawing"**. Pick the *Save All* button since the part has changed as well.

17. ▣ Add a flatness of .001 inches to the bottom surface and mark it as Datum A.

18. ▣ The outside surface of the part needs to be perpendicular to the bottom within .001 inches and is considered Datum B.

19. ▣ The hole pattern needs to be positioned within a diameter of .002 inches at maximum material condition (MMC), relative to Datum A first, then Datum B.

20. ▣ Also, the six holes need to be perpendicular to the bottom surface within a diameter of .001 inches at MMC. This becomes Datum C.

21. ▣ Make the 6.00-inch bolt circle dimension and the 60-degree angle dimension basic dimensions.

22. Save the drawing. (Pick the *Save All* button if the part has changed as well.)

23. **File> Print...** (if asked to)

24. **File> Close**

Your drawing should look similar to the following.

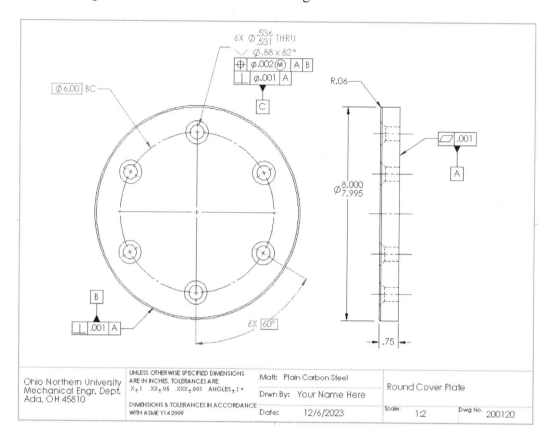

Tolerancing and GD&T Problems

10-1   Draw the symmetrical "**Table Block**" made from plain carbon steel shown below.
Create a 3-view (Front, Right, and Top) engineering drawing at ½-scale, drawing
number 121001. The <u>bottom of the table block</u> needs to be flat within 0.003 inches
and <u>will become Datum A</u>. The front surface of the table block is Datum B and it
needs to be perpendicular to the bottom within 0.004 inches. The left side in the
figure below is Datum C and it must be perpendicular to Datum A, then Datum B
within 0.004 inches.

The 1.00-inch hole goes all the way through the block and must be located within
a diameter of 0.003 inches relative to Datum A, then Datum B, and finally Datum
C. Convert the 3.25-inch from the left side and 1.25-inch from the front locational
dimensions to basic dimensions. Also, the hole's center must be perpendicular to
Datum A within a diameter of 0.001 inches at MMC. The 1.00-inch hole has an H7
fit, which means it has an upper limit of 1.0008 and a lower limit of 1.0000 inches.

Be sure to indicate which dimensions are basic dimensions.

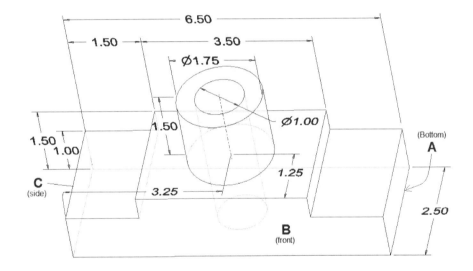

10-2   Create the "**Support Bracket**" made from AISI 1020 steel, then create a 2-view engineering drawing at 0.75-scale, drawing number 122002. Datums A, B, and C are shown below. Datum A must be flat within 0.004 inches. Datum B must be perpendicular to Datum A within 0.004 inches. Datum C must be perpendicular to Datum A, then Datum B within 0.004 inches.

The 5/8-inch diameter holes have an upper limit of 0.635 and a lower limit of 0.628 inches. The two 5/8-inch holes must be located within a diameter of 0.002 inches relative to their true position at their maximum material condition based upon Datums A, B, and C in that order.

The ¾-inch diameter hole has an upper limit of 0.760 and a lower limit of 0.754 inches. The positional tolerance for the ¾-inch hole is .002 inches in diameter at MMC relative to Datums C, B, then A in that order.

Be sure to indicate which dimensions are basic dimensions. Remember that basic dimensions are considered exact values and the GD&T affects how far the feature can vary from its exact value.

Be sure Datum tag A and Datum tag C do not line up with the 1.50-inch dimensions. If a datum tag lines up with a dimension line, then the datum is considered to be at the midpoint of the dimensioned feature.

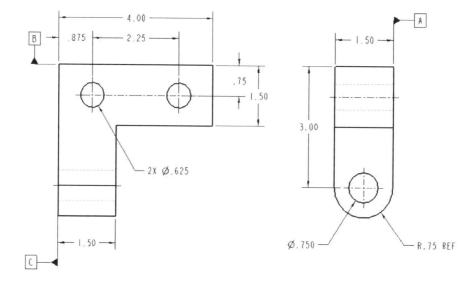

10-3   Draw the "**Spacer**" made from plain carbon steel shown below. Create a 2-view engineering drawing at ½-scale, drawing number 123003. The back of the spacer must be flat within 0.001 inches and will become Datum A. The front surface must be parallel to the back surface within 0.001 inches. The left ¾-inch hole is Datum B. It needs to be perpendicular to the back surface within a 0.001-inch diameter at maximum material condition (MMC). The right side slot is Datum C and it must be positioned within 0.003 inches relative to Datum A, then Datum B at MMC. Use the profile symbol for this GD&T feature.

The center hole must be positioned relative to Datums A, B at MMC, and C at MMC in that order within a 0.002-inch diameter at MMC. The outside surface must have a profile tolerance of 0.005 inches all around.

The ¾-inch diameter hole and slot have an upper limit of 0.760 and a lower limit of 0.754 inches. The 2¼-inch diameter hole has an upper limit of 2.265 and a lower limit of 2.250 inches. The thickness of the spacer must be between 0.378 and 0.375 inches.

Be sure to indicate which dimensions are basic dimensions. Remember that basic dimensions are considered exact values and the GD&T affects how far the feature can vary from its exact value.

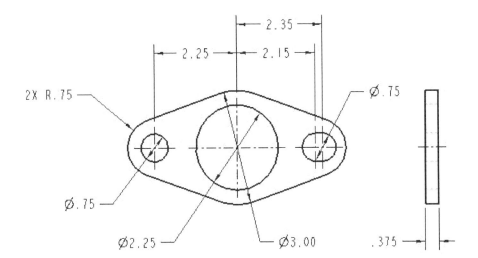

10-4   Create a 2-view engineering drawing of the **"Flanged Bushing"** made from manganese bronze at ¾-scale, drawing number 124004, as shown below. The axis of the 1-3/8-inch diameter hole's axis is Datum A so be sure to line up Datum tag A with the 1.375 dimension line. The 1 7/8-inch outside diameter must have a total runout tolerance of 0.0004 inches relative to Datum A. The right side of the flange is Datum B and must be perpendicular to Datum axis A within 0.002 inches. The left outside surface of the flange must be parallel to Datum B within 0.002 inches.

The 1-3/8-inch diameter hole has an upper limit of 1.3758 and a lower limit of 1.3750 inches. The 1-7/8-inch outside diameter of the tube has an upper limit of 1.8758 and a lower limit of 1.8753 inches. The outer diameter of the ring on the left end has an upper limit of 2.758 and a lower limit of 2.750 inches. The thickness of the ring must be between .250 and .256 inches. The overall length of the 1-7/8-inch diameter tube including the round must be between 3.000 and 2.996 inches.

Design change: The round fillet between the 2¾-inch diameter and the 1-7/8-inch diameter needs to be increased to 0.15 inches.

Be sure to indicate which dimensions are basic dimensions. Remember that basic dimensions are considered exact values and the GD&T affects how far the feature can vary from its exact value.

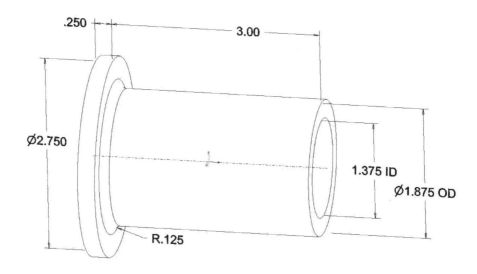

10-5   Create a 2-view engineering drawing at ½-scale for the **"Slotted Hub"** shown below, drawing number 125005. This part is to be made from AISI 1045 cold-drawn steel. The centerline of the 1.50-inch diameter hole is the Datum A with an upper limit of 1.5010 and a lower limit of 1.5000 inches. A .06 x 45° chamfer must be added to both ends on this 1.50-inch diameter hole. Section the RIGHT view to show this.

The front face (marked B in the figure below) must have a total runout geometric tolerance of 0.0005 inches relative to Datum A and is labeled Datum B. The back surface must be parallel to Datum B within 0.001 inches. The 3-inch diameter hub has an upper limit of 3.0009 and a lower limit of 3.0003 inches. This 3-inch diameter also has a total runout geometric tolerance of 0.0004 inches relative to Datum A. The ¾-inch high slot has an upper limit of 0.754 and a lower limit of 0.750 inches. It has a geometric positional tolerance of 0 at maximum material condition (MMC) and is relative to Datum A at MMC and Datum B. The center of the ¾-inch high slot then becomes Datum C so align the Datum tag C with the .75-inch dimension.

A geometric profile tolerance of 0.004 inches applies to the back side of the slot and is relative to Datum A, Datum B, and Datum C in that order. With that said, the 2.50-inch dimension becomes a basic dimension.

The three ½-inch holes have an upper limit of 0.525 and a lower limit of 0.510 inches. The three holes must be positioned within a 0.003-inch diameter of their true position relative to Datums A, B, and C in that order. With that said, the 60° angle and the 4¼-inch bolt circle become basic dimensions.

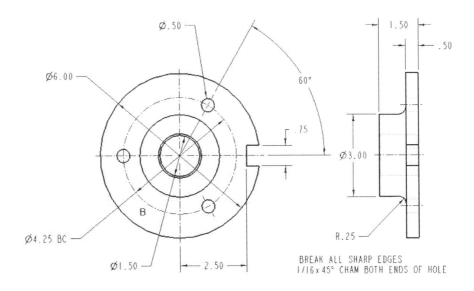

10-6   Determine the number of orthographic views necessary for the 1/8-inch thick "**Roller Bracket**" shown below, and then create the appropriate engineering drawing at full scale, drawing number 126006. Add a general note to break all sharp edges. This part is made from AISI 1010 hot rolled bar steel. The ¼-inch diameter hole has limits of .258 inches and .250 inches. The ½-inch diameter hole has limits of 0.504 and 0.500 inches.

The lower front surface (marked A) must be flat within 0.010 inches and is primary Datum A. The ¼-inch diameter hole must be perpendicular to Datum A within a 0.003-inch diameter and be designated as Datum B. The ¼-inch height of the slot has limits of 0.260 and 0.254 inches and is Datum C. Use a geometric profile tolerance of 0.003 inches relative to Datum A, then Datum B for the slot. The back side of the upper portion with the ½-inch diameter hole must be parallel to Datum A within 0.015 inches. The ½-inch diameter hole, located directly above the ¼-inch diameter hole, must be positioned within a diameter of 0.012 inches relative to Datums A, B, and C in that order.

With that said, the horizontal 1.625-inch, the horizontal 0.25-inch slot width, the horizontal 1.375-inch, and the vertical 1.50-inch dimensions become basic dimensions.

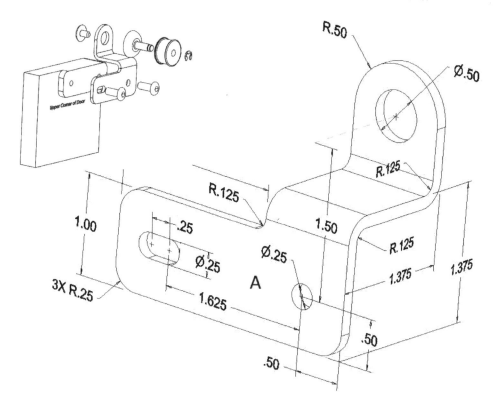

10-7   Determine the number of orthographic views necessary for the ½-inch thick "**Support Bracket_305**" shown below. It was defined in Chapter 6, problem 6-33. Create the appropriate engineering drawing at half scale, drawing number 160121. Add a general note to break all sharp edges. This part is made from cast alloy steel. The four ½-inch diameter mounting holes have limits of .530 inches and .500 inches. The two ½-inch diameter holes have limits of 0.504 and 0.500 inches. The bottom of the part needs to be flat within .003 inches and is Datum A. The lower left edge of the part is Datum B and must be perpendicular to Datum A within .005 inches. The lower left mounting hole in the Front view is Datum C. It must be perpendicular to Datum A within a diameter of .003 inches at maximum material condition. The two ½ inch diameter holes must be positioned within a diameter of .002 inches relative to Datums A and C.

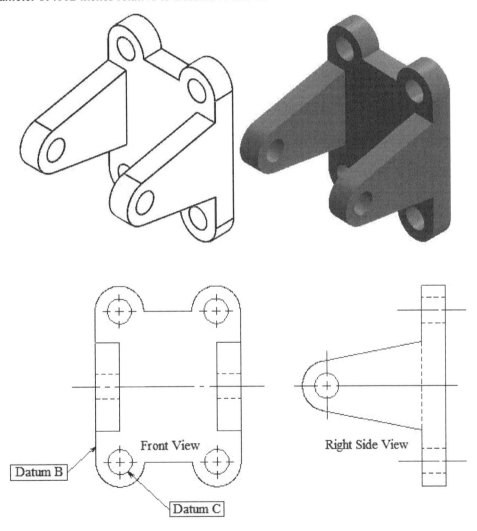

10-8   Create three views of the **"Guide"** with dimensions. This part was created in Chapter 3, problem 3-27. The sketches used to create this part are repeated here. Datum A is the bottom surface and must be flat within .001 inches. Datum B is the inner vertical portion and must be perpendicular to Datum A within .002 inches. Datum C is the left end when the part is viewed from the back. It must be perpendicular to Datum A, then Datum B within .005 inches. Datum D is the 1.125-inch diameter hole with an upper limit of 1.128 inches. It must be positioned within a diameter of .001 inches at Maximum Material Condition relative to Datum A, then Datum C. The two 21/32-inch (.656) diameter holes in the base portion of the Guide must be positioned within a diameter of .002 inches relative to Datums A, B, and D in that order. Use DrwgNo = 150100. Be sure to indicate which dimensions are basic dimensions.

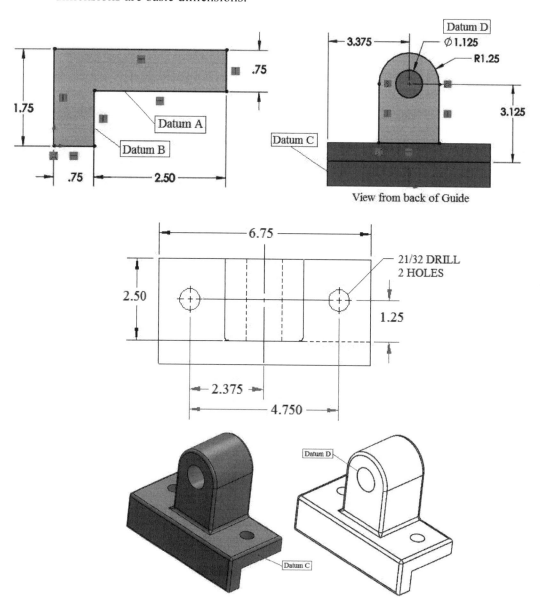

View from back of Guide

Chapter 11.　　Introduction to Finite Element Analysis

Introduction

One aspect of finite element analysis (FEA) is the numerical estimation of stresses and displacements of a 3D part by applying real-world loads and boundary conditions. This chapter focuses on an introduction to static structural analysis using SimulateXpress. This chapter will not discuss the theory behind FEA.

It is important to design a mathematical model for FEA that will reflect the actual real-world conditions. The CAD model for producing the part is not necessarily the model needed to solve for real-world stresses and displacements. Simple models for FEA work best since <u>only an estimate of the stresses and displacements is desired</u>. Complex models work best for the production of the part. Keep this in mind when creating the 3D model for FEA or when simplifying the 3D model used for the production of the part.

The FEA method breaks the 3D part into small regions and then applies the governing equations to each region. Breaking the part up into smaller regions typically gives you better results. However, having more regions also increases the computer time needed for a solution.

FEA can be divided into three categories: 1-dimensional using line elements, 2-dimensional using plane elements, and 3-dimensional using volume elements. Although 3D models can be used for all FEA problems, simplifying the model to 1D or 2D, when appropriate, reduces the computer time needed without significant loss of accuracy. Some preliminary hand calculations should be done to estimate the stresses or displacements in a few locations on the part to verify the FEA results by comparing the FEA results with the hand-calculated results.

SOLIDWORKS typically comes with two versions of FEA, Simulation, and SimulationXpress. You will use SimulationXpress in this Chapter. SOLIDWORKS lets you create a model that can be referenced by SimulationXpress. SimulationXpress utilizes ten-node tetrahedral regions with curved edges and faces. SimulationXpress solves each region for its unknown displacement in the X, Y, and Z directions. The displacement of each region is used to calculate the strain and stress at each region.

SimulationXpress is an easy-to-use application in SOLIDWORKS for a first-pass estimate of the stresses present. The Simulation application mentioned earlier will give you better results because it has more advanced capabilities.

SimulationXpress will guide you through adding material properties to the part if not already assigned, adding constraints and loads, meshing the part, solving the governing equations for stresses and displacements, and finally reviewing the results. The results section allows the user to view the stresses throughout the part, the displacement of the part in different regions, and areas where the factor of safety is less than the allowable value. The default value for the factor of safety is one.

It is best to create a separate folder different from the SolidWorks Parts folder discussed previously. This new folder can be inside the SolidWorks Parts folder or outside of it.

SimulationXpress Interface

To activate SimulationXpress, select it under the Tools tab at the top of the screen.

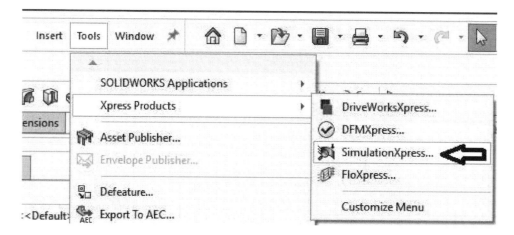

If you see the following error, don't panic. Do the following.

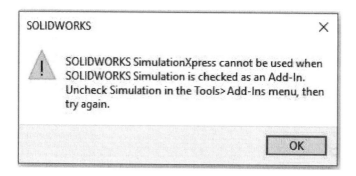

Select the Options icon at the top of the screen, then pick the **Add-Ins...** option.

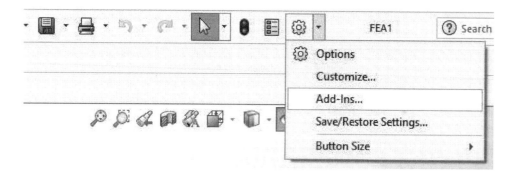

Uncheck the **SOLIDWORKS Simulation** Add-In both as an active Add-In and as an Add-In at startup, then pick OK.

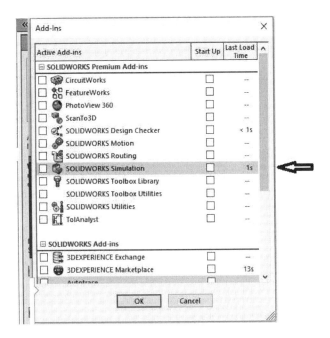

Try again to activate SimulationXpress by selecting it under the Tools tab at the top of the screen.

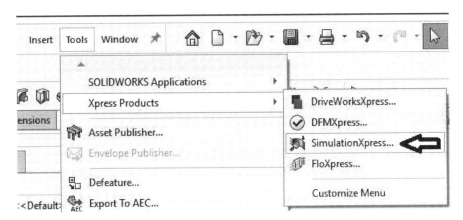

The SOLIDWORKS SimulationXpress window will appear on the right side of the graphics window. It will guide you through the six-step process.

1. Fixtures – identifies the constraints applied to the part to be analyzed.
2. Loads – applies the load to the part
3. Material – specifies the part's material needed for the strain and stress conversion.
4. Run – breaks the part into small regions, then assembles the governing equations to be solved for the unknown degrees of freedom, and then solves the system of equations.
5. Results – allows the user to view the results from the analysis such as displacement and stress, etc.
6. Optimize – an optional step to find the optimal value for a select parameter of the part such as a fillet radius or part thickness based upon specified constraints.

Finite Element Analysis Practice

Design Intent – Determine the maximum von Mises stress in an 8 in x 2 in x .50 in plain carbon steel bar with a .70-inch diameter hole located at the center of the bar when an 8000-pound axial force is applied at one end with the other end fixed. The plain carbon steel material has a Young's modulus of 3.04×10^6 psi, a Poisson's ratio of 0.28, a yield strength of 32 ksi, and an ultimate strength of 58 ksi.

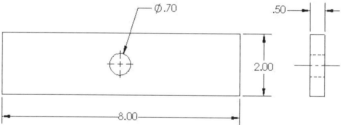

1. Before beginning this practice session, create a folder called "**FEA Practice**" either in the SolidWorks Parts folder or outside it on your desktop.
2. Start SOLIDWORKS if necessary. Be sure to toggle into **Advanced** mode. Select the **Part_IPS_ANSI** under the SOLIDWORKS Template tab, then pick OK.
3. Create the "**bar with hole**" part by sketching its shape on the Front plane, then extruding it to a .50-inch thickness.
4. Assign plain carbon steel as its material.
5. Save the part in the "**FEA Practice**" folder. Name it "**Bar With Hole**".
6. Under Tool, activate *SimulationXpress* as described on the previous page.
7. Pick *Options*, then set the System of units to *English IPS*. Pick OK. Pick *Next*.
8. Since the first item in the 6-step process is to apply constraints to the part, Pick the *Add a fixture* option.
9. Pick the left end of the bar, then pick the green checkmark.
10. Pick the *Next* option.
11. Pick the *Add a Force* option.

12. Pick the right end of the bar.
13. Make sure the Units are set to English (IPS).
14. Type 8000 for the magnitude of the force in pounds, then check the Reverse direction box so the force is pulling on the bar. Pick the green checkmark.
15. Pick the *Next* option.
16. If no part material is assigned, skip to step 18.
17. Since plain carbon steel has already been assigned, you need to verify that the units are English IPS. If the units are not Engish IPS, pick the *Change material* option.
18. Change the units to English IPS, then pick *Apply* and *Close*. Skip to step 20.
19. Pick the *Choose Material* option.
20. Pick Plain Carbon Steel from the list. Pick the Apply button, then the Close button. A Young's Modulus of 3.04579e+07 psi and Yield Strength of 31994.5 psi will appear in the window.
21. Pick the *Next* option.
22. Pick the *Run Simulation* option, then wait for the FEA analysis to finish.
23. When the analysis finishes, the result will show how the bar stretches as the force is applied. The simulation result is exaggerated. Pick the *Stop animation* option.
24. For the question, "Does the part deform as you expected?" pick *Yes, continue*.
25. On the left side of the screen, double-click on the **Stress (-vonMises-)** from the list under Results. A color-coded picture of the bar will appear along with a scale indicating the range of stresses per color. Note that the highest stress of approximately 26.8 ksi is located at the top and bottom of the hole. This is what we expected. Since the yield strength of plain carbon steel is approximately 32 ksi, the material will not yield. This means when the force is taken off the part, it will return to its original shape and length.

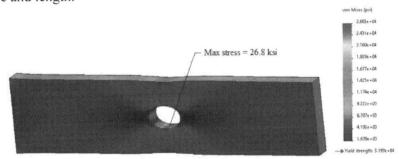

26. The change in the length of the bar under the load can be shown by double-clicking on the **Displacement (-Res disp-)** from the list under Results. The bar stretches about .00227 inches.

27. Double-clicking on the **Factor of Safety (-Max von Mises Stress-)** shows the bar in all blue which means the factor of safety is greater than 1 everywhere on the bar. The lowest factor of safety is listed in the window as 1.19 or approximately 1.2.
28. Pick the *Done viewing results* option
29. Pick the *Generate report* option, then enter a short description such as, "FEA results using default settings of SimulationXpress." Check the box in front of Designer: and enter your name.
30. Pick the *Generate* button.

SimulationXpress will generate a Microsoft Word document file with the date, your name, and the description you entered along with assumptions, model information, volumetric properties, material properties, graphic representation of where loads and fixtures were applied, mesh information, a graphic representation of the mesh used, and a graphic representation of the stresses, the displacement, and the factor of safety for the part.

31. Read through this report so you become familiar with the available results.
32. Save the report in the default folder location.
33. Pick the Generate eDrawings file option
34. Pick the Save button to save the file in the default folder location.
35. Locate the file just created and open it to view its contents, then close the file.

Now we need to check the FEA results with some hand calculations. We will use the stress concentration equation for a bar with a hole in it to estimate the maximum stress at the hole. The axial stress at the hole can be estimated by calculating the stress at the cross-section of the hole times the stress concentration factor, K_t, shown below.

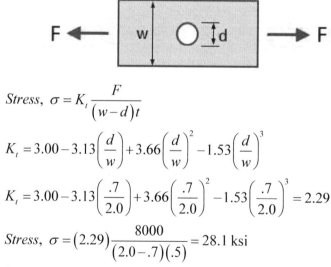

$$\text{Stress, } \sigma = K_t \frac{F}{(w-d)t}$$

$$K_t = 3.00 - 3.13\left(\frac{d}{w}\right) + 3.66\left(\frac{d}{w}\right)^2 - 1.53\left(\frac{d}{w}\right)^3$$

$$K_t = 3.00 - 3.13\left(\frac{.7}{2.0}\right) + 3.66\left(\frac{.7}{2.0}\right)^2 - 1.53\left(\frac{.7}{2.0}\right)^3 = 2.29$$

$$\text{Stress, } \sigma = (2.29)\frac{8000}{(2.0-.7)(.5)} = 28.1 \text{ ksi}$$

The FEA analysis indicated that the maximum stress was 26.8 ksi, while calculations indicate that the maximum stress is about 28.1 ksi. There is about a 5% difference in these answers but we realize that SimulationXpress is only an estimate of the stresses in the part.

If we assume the bar doesn't have a hole in it, the deflection can be calculated using:

$$\Delta L = \frac{FL}{AE} = \frac{(8000 \text{ lb.})(8.00")}{(2.00"*.5")(3.04x10^7 \text{ psi})} = .00211 \text{ inches}$$

With a hole in the bar, we would expect the change in length to be a bit longer so a change in length of .00227 inches from SimulationXpress seems reasonable.

36. Pick the *Back* option until you see the *Change setting* option.
37. Pick the *Change settings* option.
38. Pick the *Change mesh density* option.
39. When the question, "Remeshing will delete the results for study: SimulationXpress Study." Pick the OK button.

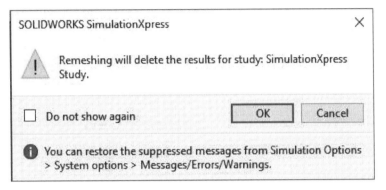

40. Slide the Mesh Density pointer to the far right which will set the mesh at Fine, then pick the green checkmark. This will remesh the part using smaller size elements.

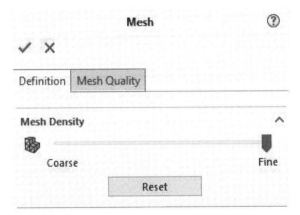

41. Pick the *Next* option.
42. Pick the *Run Simulation* option.
43. When it finishes, the animation of the elongation of the bar will begin. Pick the *Stop animation* option.
44. Pick the *Yes, continue* option.
45. Pick the *Show von Mises stress* option. Note that the maximum von Mises stress is now estimated to be 28.0 ksi which is very close to the 28.1 ksi calculated using the stress concentration factor previously. There is a difference of less than 0.4%.

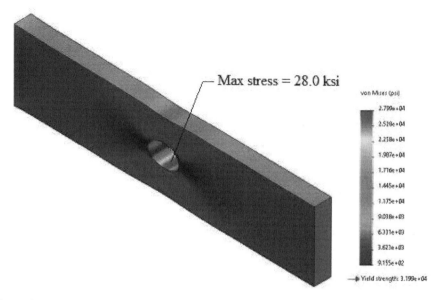

Max stress = 28.0 ksi

46. Close the SOLIDWORKS SimulationXpress window by picking the X.

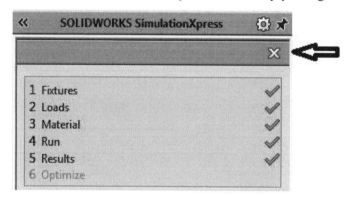

47. Pick the Yes button when asked, "Do you want to save SOLIDWORKS SimulationXpress data?"

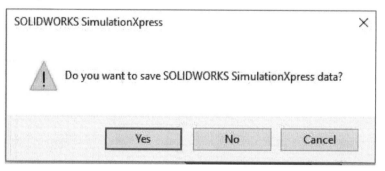

48. Save the part.

49. **File> Close**

Finite Element Analysis Exercise

Design Intent – Determine the maximum von Mises stress in a 2-inch wide, .5-inch thick, 12-inch long rectangular cantilever beam made from AISI 1010 hot-rolled bar steel when a 150-pound vertical force is applied at its right end. Assume the left end is fixed. This material has a Young's modulus of 2.90×10^6 psi, a Poisson's ratio of 0.29, a yield strength of 26 ksi, and an ultimate strength of 47 ksi.

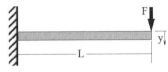

1. Before beginning, create a folder called "**FEA Exercise**" either in the SolidWorks Parts folder or outside it on your desktop.
2. Start SOLIDWORKS if necessary. Be sure to toggle into **Advanced** mode. Select the **Part_IPS_ANSI** SOLIDWORKS Template, then pick OK.
3. Create the "**cantilever beam**" part by sketching its shape on the Front plane, then extruding it to a thickness of 1.50 inches.
4. Save the part in the "**FEA Exercise**" folder. Name it "**Cantilever Beam**".
5. Activate SimulationXpress as described earlier in this chapter.
6. Pick *Options*, then set the System of units to *English IPS*. Pick OK. Pick *Next*.
7. Since the first item in the 6-step process is to apply constraints to the part, Pick the *Add a fixture* option.
8. Pick the left end of the cantilever beam, then pick the green checkmark.
9. Pick the *Next* option.
10. Pick the *Add a force* option.
11. Pick the right end of the cantilever beam.
12. Pick the *Selected direction* option, then pick the Top Plane in the FeatureManager Design Tree.
13. Make sure the Units are set to English (IPS).
14. Type 150 for the magnitude of the force in pounds, then check the Reverse direction box so the force is pushing down on the cantilever beam.

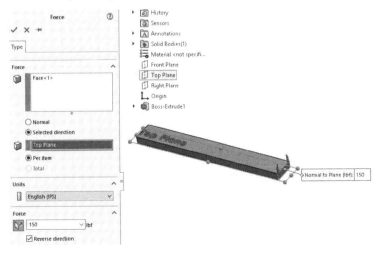

15. Pick the green checkmark.
16. Pick the *Next* option.
17. Pick the *Choose Material* option.
18. Pick AISI 1010 steel, hot rolled bar from the list. Pick the Apply button, then the Close button. A Young's Modulus of 2.90075e+07 psi and Yield Strength of 26106.8 psi will appear in the window. If metric units show up, pick the Change material option again and reselect the AISI 1010 steel, hot rolled bar material again making sure the units are set to IPS.
19. Pick the *Next* option.
20. Pick the *Run Simulation* option, then wait for the FEA analysis to finish.
21. When the analysis finishes, the result will show how the cantilever beam deflects as the vertical force is applied. The stimulation which resembles a diving board deflecting is exaggerated.
22. Pick the *Stop animation* option to stop the motion.
23. Pick *Yes, continue* for the question, "Does the part deform as you expected?"
24. On the left side of the screen, double-click on the Stress (-vonMises-) from the list under Results or pick the *Show von Mises stress* on the right. A color-coded picture of the bar will appear along with a scale indicating the range of stresses per color. Note that the highest tensile stress of approximately 28.1 ksi is located at the wall on the top of the cantilever beam. This is what we expected. Since the yield strength of plain carbon steel is approximately 26 ksi, the material will yield. This means when the force is taken off the part, it will not return to its original shape in the real world.

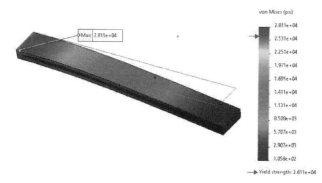

25. The vertical deflection at the end of the cantilever beam can be shown by double-clicking on the Displacement (-Res disp-) from the list under Results or by picking *Show displacement* on the right. The bar deflects vertically about .189 inches.

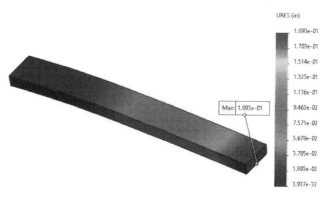

26. Double-clicking on the Factor of Safety (-Max von Mises Stress-) shows the cantilever beam has a red region near the wall which means the factor of safety is less than 1 near the wall. The lowest factor of safety is listed in the window as 0.928 or approximately 0.93.
27. Pick the *Done viewing results* option
28. Pick the *Generate report* option, then enter a short description such as, "FEA results using default settings of SimulationXpress." Check the box in front of Designer: and enter your name.
29. Pick the Generate button.

SimulationXpress will generate a Microsoft Word document file with the date, your name, and the description you entered along with assumptions, model information, volumetric properties, material properties, graphic representation of where loads and fixtures were applied, mesh information, a graphic representation of the mesh used, and a graphic representation of the stresses, the displacement, and the factor of safety for the part.

30. Read through this report so you become familiar with the available results.
31. Save the report in the default folder location.
32. Pick the *Generate eDrawings file* option
33. Pick the Save button to save the file in the default folder location.
34. Locate the file just created and open it to view its contents, then close the file.

Now we need to check the FEA results with some hand calculations. We will use the stress concentration equation for a bar with a hole in it to estimate the maximum stress at the hole. The bending stress at the wall can be estimated by calculating using the following equation.

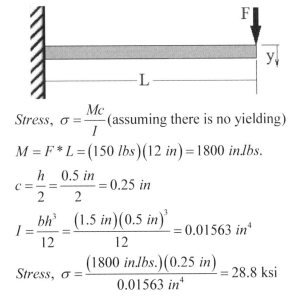

$$\text{Stress, } \sigma = \frac{Mc}{I} \text{ (assuming there is no yielding)}$$

$$M = F*L = (150 \; lbs)(12 \; in) = 1800 \; in.lbs.$$

$$c = \frac{h}{2} = \frac{0.5 \; in}{2} = 0.25 \; in$$

$$I = \frac{bh^3}{12} = \frac{(1.5 \; in)(0.5 \; in)^3}{12} = 0.01563 \; in^4$$

$$\text{Stress, } \sigma = \frac{(1800 \; in.lbs.)(0.25 \; in)}{0.01563 \; in^4} = 28.8 \; ksi$$

The FEA analysis indicated that the maximum stress was 28.1 ksi, while calculations indicate that the maximum stress is about 28.8 ksi. There is about a 2.5% difference in these answers but we realize that SimulationXpress is only an estimate of the stresses at the wall in the cantilever beam.

The deflection of the cantilever beam can be calculated using:

$$\Delta L = \frac{FL^3}{3EI} = \frac{(150 \text{ lb.})(12.00 \text{ in})^3}{3(2.90x10^7 \text{ psi})(0.01563 \text{ in}^4)} = .191 \text{ inches}$$

We would expect the vertical deflection of the cantilever beam to be about the same as the result from SimulationXpress. The results vary by less than 1%.

35. Pick the *Back* option until you see the Change setting option.
36. Pick the *Change settings* option.
37. Pick the *Change mesh density* option.
38. When the question, "Remeshing will delete the results for study: SimulationXpress Study." Pick the OK button.

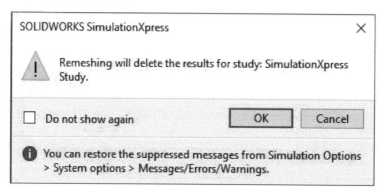

39. Slide the Mesh Density pointer to the far right which will set the mesh at Fine, then pick the green checkmark. This will remesh the part using smaller size elements.

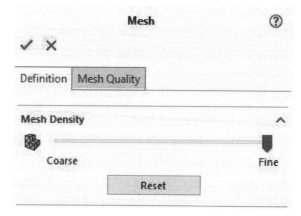

40. Pick the *Next* option.
41. Pick the *Run Simulation* option.
42. When it finishes, the animation of the elongation of the bar will begin. Pick the *Stop animation* option.
43. Pick the *Yes, continue* option.

44. Pick the *Show von Mises stress* option. Note that the maximum von Mises stress is still estimated to be 28.1 ksi. There is a difference is still less than 0.4%. Because the material yielded, the stress estimated by SimulationXpress will be slightly less. The hand calculation for stress assumes no yielding of the material occurs.
45. Pick the *Done viewing results* option.
46. Close the *SOLIDWORKS SimulationXpress* window by picking the X.

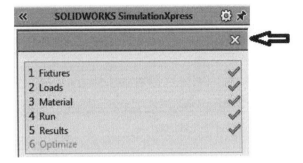

47. Pick the Yes button when asked, "Do you want to save SOLIDWORKS SimulationXpress data?"

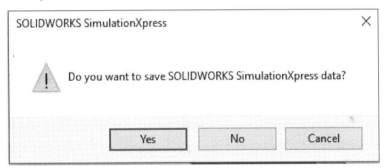

48. ⊟ Save the part.
49. **File> Close**

Finite Element Analysis Problems

11-1 Before you begin, create a folder called **"FEA Shaft Design"**, then save the part in this folder. Name the part **"Shaft Design"**. Design the AISI 1045 CD steel, constant diameter shaft shown below, given the approximate constant loading shown. Assume the bearings at O and B are 3.00 inches in diameter and fixed. Initial estimates led the design engineer to a 1.50-inch diameter shaft. However, his supervisor stated that the deflection at the big pulley must be less than .025 inches. At points A and C combine the loads and then apply the combined load to the appropriate pulley surface. Both pulleys and bearings are 2.00 inches wide and may be modeled as a solid disk. Use SimulationXpress to determine the deflection of the steel shaft at point C. If the deflection requirement (less than .025 inches) is not met, then increase the shaft's diameter in 0.250-inch increments until the deflection requirement is met. We are not concerned with the twist or the rotation of the shaft at this time. What is the recommended standard shaft diameter so the deflection at point C is less than .025 inches? Diameter = _____ inches. (Note: With a 1.50-inch diameter shaft, the deflection at point C is approximately .045 inches > .025 inches.)

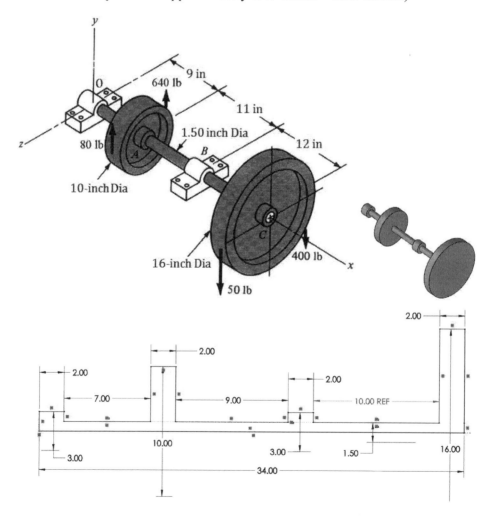

11-2   Before you begin, create a folder called **"FEA Lever Design"**, then save the part in this folder. Name the part **"Lever Design"**. The tapered lever shown below needs to be redesigned to use less Galvanized steel by decreasing the radius of the right end of the lever, thus lowering the lever's cost. The hole at the left end is 1.50 inches in diameter with a slot cut for a square key. The square key is 3/8" x 3/8" x 1.00" and the distance to the top of the keyway from the opposite side of the hole is 1.670 inches. The outer diameter at the left end is 3.00 inches. The lever is 1.00 inches thick. A maximum force of 300 pounds is applied to the last, flat 2.00 inches of the 15-inch lever (center of the hole to far right end). The height of the lever near the hub is 2.00 inches. Using a safety factor of 4.5 for Galvanized steel, the maximum allowable von Mises stress is 6580 psi. (Answer: R = 0.265 in)

What is the weight savings for the redesigned tool? Initial weight = _____ lbs. Final weight _____ lbs. Percent savings in weight _____%.

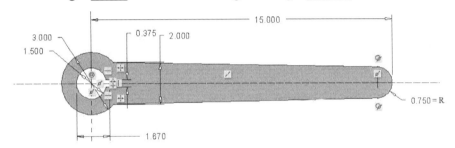

Add a 2.00-inch radius fillet after extruding the Lever.

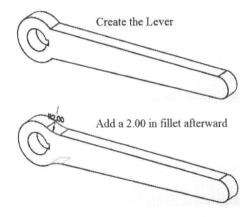

Create the Lever

Add a 2.00 in fillet afterward

The 300-lb force is applied to the last 2.00 inches of the flat part of the handle. You may need to create a surface patch .001 inches thick, then apply the vertical force.

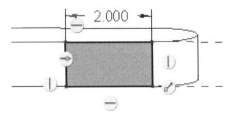

11-3 Before you begin, create a folder called **"FEA Hanger"**, then save the part in this folder. Name the part **"Hanger"**. Perform a stress and defection analysis of the Hanger shown below. The hanger is made from AISI 1045 CD steel. A 4.0-pound vertical downward force is applied in the .200-inch radius channel centered 4.10 inches from the wall. Use the Top plane as the direction for the vertical force. The hanger is attached to the wall using two 1/4-20 UNC cap screws. Assume the two holes in the hanger are fixed at the wall. Determine the maximum von Mises stress _____ psi in the hanger and its location. Determine the maximum vertical deflection of the hanger _____ inches and its location. Assuming a safety factor of at least 4, is the AISI 1045 CD steel sufficient? (Yes/No) The safety factor is _____ .

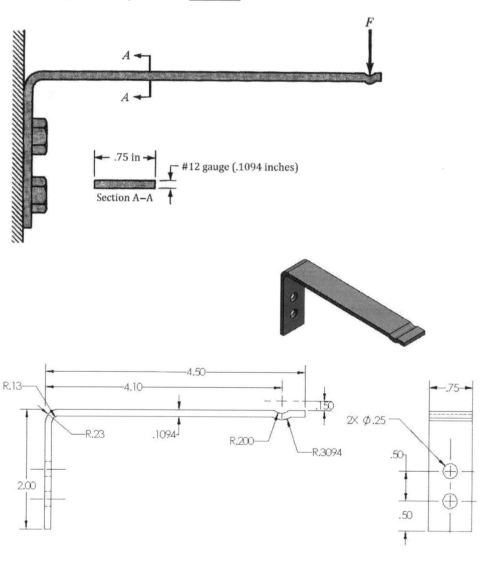

11-4 Before you begin, create a folder called **"FEA C-Clamp"**, then save the part in this folder. Name the part **"C-Clamp"**. The C-clamp is made from AISI 1010 hot-rolled bar steel with a yield strength of approximately 26.1 ksi. A 100-lb. force is applied in the top hole. The bottom hole is fixed. Note: The top and bottom holes of the C-clamp must be created separately because they have different conditions applied during FEA. Determine the maximum von Mises stress in the C-clamp _____ psi and its location. Determine the vertical deflection of the top hole _____ inches. Based upon 26.1 ksi, what is the safety factor for the C-clamp? _____

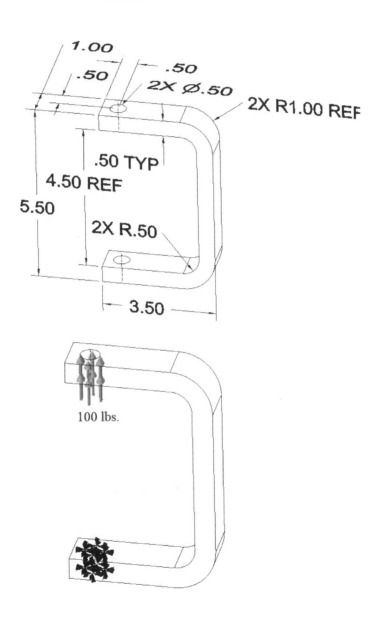

100 lbs.

11-5 Before you begin, create a folder called **"FEA Crank"**, then save the part in this folder. Name the part **"Crank"**. A 25-mm diameter, solid steel rod is firmly attached to a wall as shown below. The inner edge of the crank lever is located 150 mm from the wall. The center of the handle is located 125 mm from the center of the 25-mm diameter rod. Place 3-mm rounds where the handle meets the solid rods to simulate welds. A force of (50, -800, 300) N is applied to the end of the crank. In SimulationXpress apply three different forces to the end of the crank, one in each direction. Use the Top plane direction for the 800 N force, and the Right plane for the 300 N force. Assume the crank is made from AISI 1020 cold-rolled steel. Be sure to change the units to MMGS before creating the crank. <u>Be sure to change the units to **SI** before running the SimulationXpress analysis.</u>

   a.   What is the magnitude of the deflection at the far end of the handle?
        _____ mm
   b.   What is the maximum von Mises stress at the wall? _____ MPa
   c.   What is the minimum factor of safety? _____
   d.   Does the FEA analysis agree with the closed-form equations that you learned in the mechanical design course? (Yes/No)
   e.   Calculate the maximum normal bending stress at the wall. Calculate the maximum shear stress at the wall. Then calculate the maximum von Mises stress at the wall. _____ MPa

$$\sigma'_{vonMises} = \sqrt{\sigma_N^2 + 3\tau_{xy}^2}$$

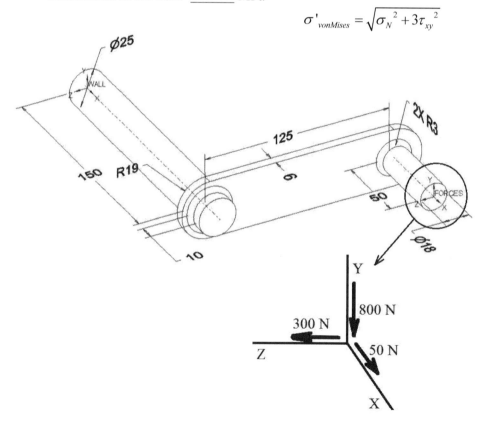

Chapter 12. Appendices

Appendix A – Drill and Tap Chart

Appendix B – Number and Letter Drill Sizes

Appendix C – Surface Roughness

Appendix D – Clevis Pin Sizes

Appendix E – Square and Flat Key Sizes

Appendix F – Screw Sizes

Appendix G – Nut Sizes

Appendix H – Setscrew Sizes

Appendix I – Washer Sizes

Appendix J – Basic Hole Tolerance

Appendix K – Basic Shaft Tolerance

Appendix L – Inch Tolerance Zones

Appendix M – Metric Tolerance Zones

Appendix N – International Tolerance Zones

Appendix A – Drill and Tap Chart

Major Dia		Threads (Thd/in)	Tap Drill Size		Clearance Hole	
Size	(inch)		Drill	Decimal	Drill	Decimal
0	0.06	80	3/64	0.0469	50	0.07
1	0.073	64	53	0.0595	46	0.081
		72	53	0.0595		
2	0.086	56	50	0.07	41	0.096
		64	50	0.07		
3	0.099	48	47	0.0785	35	0.11
		56	45	0.082		
4	0.112	40	43	0.089	30	0.1285
		48	42	0.0935		
5	0.125	40	38	0.1015	29	0.136
		44	37	0.104		
6	0.138	32	36	0.1065	25	0.1495
		40	33	0.113		
8	0.164	32	29	0.136	16	0.177
		36	29	0.136		
10	0.19	24	25	0.1495	7	0.201
		32	21	0.159		
12	0.216	24	16	0.177	1	0.228
		28	14	0.182		
		32	13	0.185		
1/4	0.25	20	7	0.201	H	0.266
		28	3	0.213		
		32	7/32	0.2188		
5/16	0.3125	18	F	0.257	Q	0.332
		24	I	0.272		
		32	9/32	0.2812		

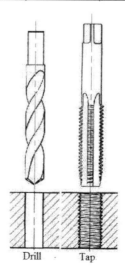

Drill   Tap

Major Dia		Threads (Thd/in)	Tap Drill Size		Clearance Hole	
Size	(inch)		Drill	Decimal	Drill	Decimal
3/8	0.375	16	5/16	0.3125	X	0.397
		24	Q	0.332		
		32	11/32	0.3438		
7/16	0.4375	14	U	0.368	15/32	0.4687
		20	25/64	0.3906		
		28	Y	0.404		
1/2	0.5	13	27/64	0.4219	17/32	0.5312
		20	29/64	0.4531		
		28	15/32	0.4688		
9/16	0.5625	12	31/64	0.4844	19/32	0.5938
		18	33/64	0.5156		
		24	33/64	0.5156		
5/8	0.625	11	17/32	0.5312	21/32	0.6562
		18	37/64	0.5781		
		24	37/64	0.5781		
11/16	0.6875	24	41/64	0.6406	23/32	0.6562
3/4	0.75	10	21/32	0.6562	25/32	0.7812
		16	11/16	0.6875		
		20	45/64	0.7031		
13/16	0.8125	20	49/64	0.7656	27/32	0.8438
7/8	0.875	9	49/64	0.7656	29/32	0.9062
		14	13/16	0.8125		
		20	53/64	0.8281		
15/16	0.9375	20	57/64	0.8906	31/32	0.9688
1	1	8	7/8	0.875	1 1/32	1.0313
		12	15/16	0.9375		
		20	61/64	0.9531		

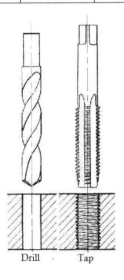

Drill   Tap

Appendix B – Number and Letter Drill Sizes

Drill Size Number	Diameter (inches)	Diameter (mm)	Drill Size Number	Diameter (inches)	Diameter (mm)	Drill Size Number	Diameter (inches)	Diameter (mm)
1	0.228	5.79	28	0.141	3.58	55	0.052	1.32
2	0.221	5.61	29	0.136	3.45	56	0.047	1.19
3	0.213	5.41	30	0.129	3.28	57	0.043	1.09
4	0.209	5.31	31	0.120	3.05	58	0.042	1.07
5	0.206	5.23	32	0.116	2.95	59	0.041	1.04
6	0.204	5.18	33	0.113	2.87	60	0.040	1.02
7	0.201	5.11	34	0.111	2.82	61	0.039	0.99
8	0.199	5.05	35	0.110	2.79	62	0.038	0.97
9	0.196	4.98	36	0.107	2.72	63	0.037	0.94
10	0.194	4.93	37	0.104	2.64	64	0.036	0.91
11	0.191	4.85	38	0.102	2.59	65	0.035	0.89
12	0.189	4.80	39	0.100	2.54	66	0.033	0.84
13	0.185	4.70	40	0.098	2.49	67	0.032	0.81
14	0.182	4.62	41	0.096	2.44	68	0.031	0.79
15	0.180	4.57	42	0.094	2.39	69	0.029	0.74
16	0.177	4.50	43	0.089	2.26	70	0.028	0.71
17	0.173	4.39	44	0.086	2.18	71	0.026	0.66
18	0.170	4.32	45	0.082	2.08	72	0.025	0.64
19	0.166	4.22	46	0.081	2.06	73	0.024	0.61
20	0.161	4.09	47	0.079	2.01	74	0.023	0.58
21	0.159	4.04	48	0.076	1.93	75	0.021	0.53
22	0.157	3.99	49	0.073	1.85	76	0.020	0.51
23	0.154	3.91	50	0.070	1.78	77	0.018	0.46
24	0.152	3.86	51	0.067	1.70	78	0.016	0.41
25	0.150	3.81	52	0.064	1.63	79	0.015	0.38
26	0.147	3.73	53	0.060	1.52	80	0.014	0.36
27	0.144	3.66	54	0.055	1.40			

Drill Size Letter	Diameter (inches)	Diameter (mm)	Drill Size Letter	Diameter (inches)	Diameter (mm)	Drill Size Letter	Diameter (inches)	Diameter (mm)
A	0.234	5.94	J	0.277	7.04	S	0.348	8.84
B	0.238	6.05	K	0.281	7.14	T	0.358	9.09
C	0.242	6.15	L	0.290	7.37	U	0.368	9.35
D	0.246	6.25	M	0.295	7.49	V	0.377	9.58
E	0.250	6.35	N	0.302	7.67	W	0.386	9.80
F	0.257	6.53	O	0.316	8.03	X	0.397	10.08
G	0.261	6.63	P	0.323	8.20	Y	0.404	10.26
H	0.266	6.76	Q	0.332	8.43	Z	0.413	10.49
I	0.272	6.91	R	0.339	8.61			

Appendix C – Surface Roughness Chart

Surface Finish micro inches	2000	1000	500	250	125	63	32	16	8	4	2	1
Metal Cutting												
Sawing Planing,	▒	■	■	■	■	▒						
Shaping Drilling		▒	■	■	■	▒	▒					
Milling			▒	■	■	■						
Boring, Turning		▒	▒	■	■	■	▒					
Broaching				■	■	■	■	▒	▒	▒	▒	▒
Reaming				▒	■	■	▒					
Forming	▒	■										
Hot Rolling		▒	■	■	▒							
Forging			▒	■	■							
Extruding			▒	▒	■	■						
Cold Rolling, Drawing				▒	■	■	▒					
Roller Burnishing							▒	■	▒			
Miscellaneous	▒	■	▒									
Flame Cutting			▒	■	■	▒						
Chemical Milling				■	■	■	▒					
Electron Beam Cutting				■	■	■	▒					
Laser Cutting				■	■	■	▒					
EDM			▒	■	■	▒						
Abrasive					▒	■	■	■	■	▒	▒	
Grinding Barrel						▒	■	■	■	▒	▒	
Finishing						▒	■	■	■	■	▒	▒
Honing						▒	■	■	■	■	▒	▒
Electro-polishing							▒	■	■	■	▒	▒
Electrolytic Grinding							▒	■	■	■	▒	▒
Polishing							▒	■	■	■	■	▒
Lapping							▒	■	■	■	■	▒
Superfinishing								▒	■	■	■	▒
Surface Finish μ-inches	2000	1000	500	250	125	63	32	16	8	4	2	1

Appendix D – Clevis Pin Sizes

Pin Dia (inch) A	Head Dia (inches) B	Hd Height (inches) C	Min. Length (inches) D	Pin Locate (inches) E	Hole Size (inches) F
0.188	0.31	0.06	0.59	0.11	0.078
0.250	0.38	0.09	0.80	0.12	0.078
0.312	0.44	0.09	0.97	0.16	0.109
0.375	0.50	0.12	1.09	0.16	0.109
0.500	0.62	0.16	1.42	0.22	0.141
0.625	0.81	0.20	1.72	0.25	0.141
0.750	0.94	0.25	2.05	0.30	0.172
1.000	1.19	0.34	2.62	0.36	0.172

Pin Dia (mm) A	Head Dia (mm) B	Hd Height (mm) C	Min. Length (mm) D	Pin Locate (mm) E	Hole Size (mm) F
4	6	1	16	2.2	1
6	10	2	20	3.2	1.6
8	14	3	24	3.5	2
10	18	4	28	4.5	3.2
12	20	4	36	5.5	3.2
16	25	4.5	44	6	4
20	30	5	52	8	5
24	36	6	66	9	6.3

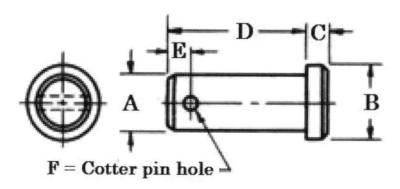

F = Cotter pin hole

Appendix E – Square and Flat Key Sizes

Shaft Diameter		Key Size		Key Size Depth (inches)
Over (inch)	Including (inch)	w (inch)	h (inch)	
5/16	7/16	3/32	3/32	3/64
7/16	9/16	1/8	3/32	3/64
		1/8	1/8	1/16
9/16	7/8	3/16	1/8	1/16
		3/16	3/16	3/32
7/8	1 1/4	1/4	3/16	3/32
		1/4	1/4	1/8
1 1/4	1 3/8	5/16	1/4	1/8
		5/16	5/16	5/32
1 3/8	1 3/4	3/8	1/4	1/8
		3/8	3/8	3/16
1 3/4	2 1/4	1/2	3/8	3/16
		1/2	1/2	1/4
2 1/4	2 3/4	5/8	7/16	7/32
		5/8	5/8	5/16
2 3/4	3 1/4	3/4	1/2	1/4
		3/4	3/4	3/8
3 1/4	3 3/4	7/8	7/8	7/16
3 3/4	4 1/2	1	1	1/2

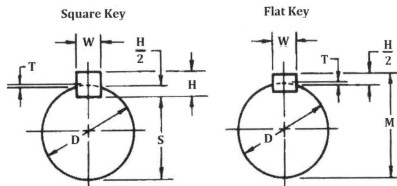

Square Key **Flat Key**

C = Allowance for parallel keys = .005 inches
W = Nominal key width (Inches)

$$S = D - \frac{H}{2} - T = \frac{D - H + \sqrt{D^2 - W^2}}{2} \qquad T = \frac{D - \sqrt{D^2 - W^2}}{2}$$

$$M = D - T + \frac{H}{2} + C = \frac{D + H + \sqrt{D^2 - W^2}}{2} + C$$

Shaft Diameter		Key Size		Key Size Depth (mm)
Over (mm)	Including (mm)	w (mm)	h (mm)	
6	8	2	2	1.0
8	10	3	3	1.5
10	12	4	4	2.0
12	17	5	5	2.5
17	22	6	6	3.0
22	30	7	7	3.5
		8	7	3.5
30	38	8	8	4.0
		10	8	4.0
38	44	9	9	4.5
		12	8	4.0
44	50	10	10	5.0
		14	9	4.5
50	58	12	12	6.0
		16	10	5.0

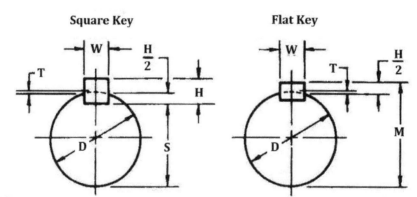

Square Key   **Flat Key**

C = Allowance for parallel keys = 0.12 mm.
W = Nominal key width (millimeters)

$$S = D - \frac{H}{2} - T = \frac{D - H + \sqrt{D^2 - W^2}}{2} \qquad T = \frac{D - \sqrt{D^2 - W^2}}{2}$$

$$M = D - T + \frac{H}{2} + C = \frac{D + H + \sqrt{D^2 - W^2}}{2} + C$$

Appendix F – Screw Sizes

Nominal Size (inch)	Slot Width (inches)	Flat & Oval Head		Round Head		Hexagon Head	
		A (inches)	B (inches)	C (inches)	D (inches)	E (inches)	F (inches)
0 (.060)	0.023	0.119	0.035	0.113	0.053	-	-
1 (.073)	0.026	0.146	0.043	0.138	0.061	-	-
2 (.086)	0.031	0.172	0.051	0.162	0.069	0.125	0.050
3 (.099)	0.035	0.199	0.059	0.187	0.078	0.187	0.055
4 (.112)	0.039	0.225	0.067	0.211	0.086	0.187	0.060
5 (.125)	0.043	0.252	0.075	0.236	0.095	0.187	0.070
6 (.138)	0.048	0.279	0.083	0.260	0.103	0.250	0.080
8 (.164)	0.054	0.332	0.100	0.309	0.120	0.250	0.110
10 (.190)	0.06	0.385	0.116	0.359	0.137	0.312	0.120
12 (.216)	0.067	0.438	0.132	0.408	0.153	0.312	0.155
1/4	0.075	0.507	0.153	0.472	0.175	7/16	0.172
5/16	0.084	0.635	0.191	0.590	0.216	1/2	0.219
3/8	0.094	0.762	0.230	0.708	0.256	9/16	0.250
7/16	0.094	0.812	0.223	0.750	0.328	5/8	0.297
1/2	0.106	0.875	0.223	0.813	0.355	3/4	0.344
9/16	0.118	1	0.260	0.938	0.410	13/16	0.359
5/8	0.133	1.125	0.298	1.000	0.438	15/16	0.422
3/4	0.149	1.375	0.372	1.250	0.547	1 1/8	0.500

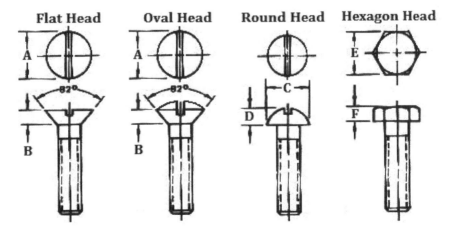

Flat Head **Oval Head** **Round Head** **Hexagon Head**

Nominal Size (mm)	Slot Width (mm)	Flat & Oval Head		Round Head		Hexagon Head	
		A (mm)	B (mm)	C (mm)	D (mm)	E (mm)	F (mm)
3	1	5.6	1.6	6	2.4	5.5	2
4	1.3	7.5	2.2	8	3.2	7	2.8
5	1.5	9.2	2.5	9.8	4	8.5	3.5
6	1.7	11	3	11.8	4.7	10	4
8	2.1	14	4	14.8	6	13	5.5
10	2.6	18	5	19.2	7.6	15	7
12	3	23	6.4	24.5	9.7	18	8
14	3.2	26	7.2	27.8	11	21	9.3
16	3.2	29	8	30.8	12.2	24	10.5
20	4.2	35	9	37.2	14.8	30	13.1
24	4.8	41	11.6	43.5	17	36	15.6
30	6	51	14.4	54.2	22	46	19.5
36	7	60	17	63.8	25	55	23.4

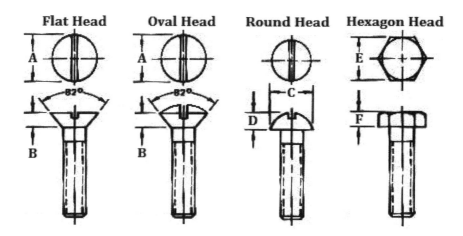

Appendix G – Nut Sizes

Major Dia Nominal Size (inch)	Distance across Flats (inches)	Height	
		Regular (inches)	Thick (inches)
1/4	7/16	7/32	9/32
5/16	1/2	17/64	21/64
3/8	9/16	21/64	13/32
7/16	11/16	3/8	29/64
1/2	3/4	7/16	9/16
9/16	7/8	31/64	39/64
5/8	15/16	35/64	23/32
3/4	1 1/8	41/64	13/16
7/8	1 5/16	3/4	29/32
1	1 1/2	55/64	1
1 1/8	1 11/16	31/32	1 5/32
1 1/4	1 7/8	1 1/16	1 1/4
1 3/8	2 1/16	1 11/16	1 3/8
1 1/2	2 1/4	1 9/32	1 1/2

Major Dia Nominal Size (mm)	Distance across Flats (mm)	Height	
		Regular (mm)	Thick (mm)
4	7	-	3.2
5	0.4	4.7	5.1
6	0.5	5.2	5.7
8	0.6	6.8	7.5
10	0.7	8.4	9.3
12	0.8	10.8	12.0
14	0.9	12.8	14.1
16	0.9	14.8	16.4
20	1.1	18.0	20.3
24	1.3	21.5	23.9
30	1.5	25.6	28.6
36	1.7	31.0	34.7

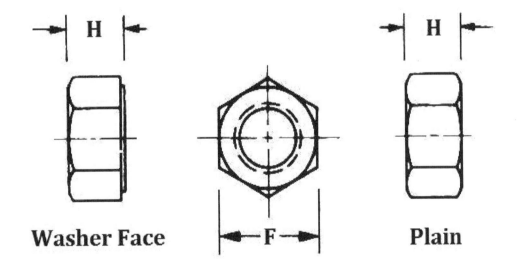

Washer Face   **F**   **Plain**

Appendix H – Setscrew Sizes

Shaft Diameter		Setscrew Diameter (inches)	Seating Torque (in.lb.)
Over (inch)	Including (inch)		
–	–	0.125	10
–	–	0.138	10
–	–	0.164	20
5/16	7/16	0.190	36
7/16	9/16	0.250	87
9/16	7/8	0.375	290
7/8	1 1/4	0.500	450
1 1/4	1 3/8	0.625	620
1 3/8	1 3/4	0.750	2400
1 3/4	2 1/4	1.000	7200
2 1/4	2 3/4	1.250	15000

Shaft Diameter		Setscrew Diameter (mm)	Seating Torque (N.m)
Over (mm)	Including (mm)		
–	–	1.4	1
–	–	2	1.1
–	–	3	2.3
6	8	4	4
8	10	6	10
10	12	8	25
12	17	10	40
17	22	12	50
22	30	14	75
30	38	16	100
38	44	18	280
44	50	20	400

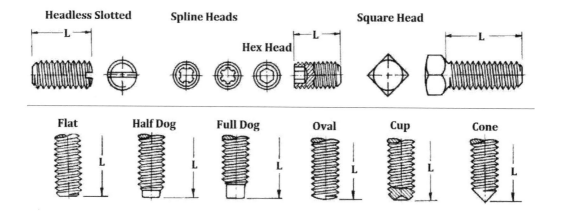

Headless Slotted   Spline Heads   Hex Head   Square Head

Flat   Half Dog   Full Dog   Oval   Cup   Cone

Appendix I – Washer Sizes

Flat Washers			
Screw Size No. or Inch	Inside Diameter (inches)	Thickness (inches)	Outside Diameter (inches)
6 (.138)	5/32	0.049	3/8
8 (.164)	3/16	0.049	7/16
10 (.190)	7/32	0.049	1/2
12 (.216)	1/4	0.065	9/16
1/4	9/32	0.065	5/8
5/16	11/32	0.065	11/16
3/8	13/32	0.065	13/16
7/16	15/32	0.065	59/64
1/2	17/32	0.095	1 1/16
9/16	19/32	0.095	1 5/32
5/8	21/32	0.095	1 5/16
3/4	13/16	0.134	1 15/32
7/8	15/16	0.134	1 3/4
1	1 1/16	0.134	2
1 1/8	1 1/4	0.134	2 1/4
1 1/4	1 3/8	0.165	2 1/2
1 3/8	1 1/2	0.165	2 3/4
1 1/2	1 5/8	0.165	3
1 3/4	1 7/8	0.180	4
2	2 1/8	0.180	4 1/2

Lock Washers			
Screw Size No. or Inch	Inside Diameter (inches)	Thickness (inches)	Outside Diameter (inches)
4 (.112)	0.120	0.025	0.209
5 (.125)	0.133	0.031	0.236
6 (.138)	0.148	0.031	0.250
8 (.164)	0.174	0.040	0.293
10 (.190)	0.200	0.047	0.334
12 (.216)	0.227	0.056	0.377
1/4	0.262	0.062	0.489
5/16	0.320	0.078	0.586
3/8	0.390	0.094	0.683
7/16	0.455	0.109	0.779
1/2	0.518	0.125	0.873
9/16	0.582	0.141	0.971
5/8	0.650	0.156	1.079
3/4	0.770	0.188	1.271
7/8	0.905	0.219	1.464
1	1.042	0.250	1.661
1 1/8	1.172	0.281	1.853
1 1/4	1.302	0.312	2.045
1 3/8	1.432	0.344	2.239
1 1/2	1.561	0.375	2.430

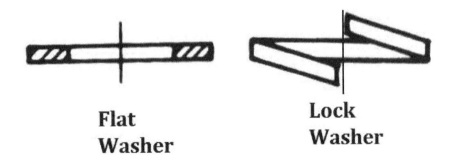

Flat Washer　　　　　**Lock Washer**

Flat Washers			
Screw Size (mm)	Inside Diameter (mm)	Thickness (mm)	Outside Diameter (mm)
2	2.5	0.35	5
3	3.5	0.55	7
4	4.7	0.9	9
5	5.5	1.1	10
6	6.6	1.8	12
8	8.9	1.8	16
10	10.8	2.2	20
12	13.3	2.7	24
14	15.2	2.7	28
16	17.2	3.3	30
18	19.2	3.3	34
20	21.8	3.3	37
22	23.4	4.3	42
24	25.6	4.3	44
27	28.8	4.3	50
30	32.4	4.3	56
36	38.3	5.6	66

Lock Washers			
Screw Size (mm)	Inside Diameter (mm)	Thickness (mm)	Outside Diameter (mm)
2	2.1	0.5	4.4
3	3.1	0.8	6.2
4	4.1	0.9	7.6
5	5.1	1.2	9.2
6	6.1	1.6	11.8
8	8.2	2	14.8
10	10.2	2.2	18.1
12	12.2	2.5	21.1
14	14.2	3	24.1
16	16.2	3.5	27.4
18	18.2	3.5	29.4
20	20.2	4	33.6
22	22.5	4	35.9
24	24.5	5	40
27	27.5	5	43
30	30.5	6	48.2
36	33.5	6	55.2

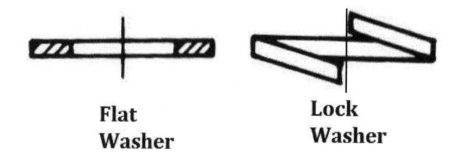

Flat Washer   **Lock Washer**

Appendix J – Retaining Ring Sizes

Nominal Shaft Diameter (inch)	Minimum Groove Diameter (inches)	Maximum Groove Diameter (inches)	Retaining Ring Diameter (inches)	Retaining Ring Thickness (inches)
1/4	0.218	0.032	0.311	0.025
5/16	0.274	0.032	0.376	0.025
3/8	0.333	0.032	0.448	0.025
1/5	0.447	0.042	0.581	0.025
5/8	0.560	0.042	0.715	0.035
3/4	0.673	0.049	0.845	0.042
7/8	0.786	0.049	0.987	0.042
1	0.897	0.049	1.127	0.042
1 1/8	1.009	0.060	1.267	0.050
1 1/4	1.122	0.060	1.410	0.050
1 3/8	1.233	0.060	1.550	0.050
1 1/2	1.346	0.060	1.691	0.050
1 3/4	1.571	0.072	1.975	0.062

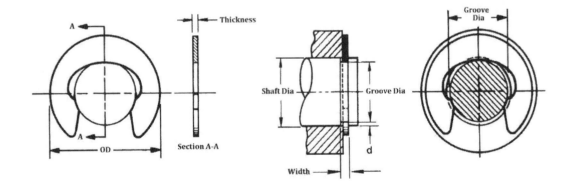

Nominal Shaft Diameter (mm)	Minimum Groove Diameter (mm)	Maximum Groove Diameter (mm)	Retaining Ring Diameter (mm)	Retaining Ring Thickness (mm)
8	6.9	0.8	10.0	0.6
10	8.9	0.8	12.2	0.6
12	10.8	0.8	14.4	0.6
14	12.6	1.2	16.3	1.0
16	14.4	1.2	18.5	1.0
18	16.2	1.2	20.4	1.2
20	17.9	1.4	22.6	1.2
22	19.7	1.4	25.0	1.2
24	21.5	1.4	27.1	1.2
25	22.4	1.4	28.3	1.2
30	26.8	1.5	33.7	1.5
35	31.3	1.8	39.4	1.5
40	25.8	1.8	45.0	1.5
45	40.3	1.8	50.6	1.5
50	44.8	2.4	56.4	2.0

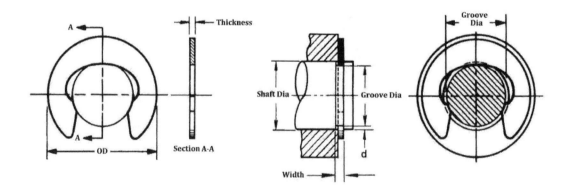

Appendix J – Basic Hole Tolerance

Basic Hole Symbol	Description of Basic Hole Fit
H11/c11	Loose running fit—for wide commercial tolerances or allowances on external members.
H9/d9	Free running fit—not for use where accuracy is essential, but good for large temperature variations, high running speeds, or heavy journal pressures.
H8/f7	Close running fit—for running on accurate machines and for accurate location at moderate speeds and journal pressures.
H7/g6	Sliding fit—not intended to run freely, but to move and turn freely and locate accurately.
H7/h6	Locational clearance fit—provides snug fit for locating stationary parts; but can be freely assembled and disassembled.
H7/k6	Location transition fit—for accurate location, a compromise between clearance and interference.
H7/n6	Locational transition fit—for more accurate location where greater interference is permissible.
H7/p6	Locational interference fit—for parts requiring rigidity and alignment with prime accuracy of location, but without special bore pressure requirements.
H7/s6	Medium drive fit—for ordinary steel parts or shrink fits on light sections, the tightest fit usable with cast iron.
H7/u6	Force fit—suitable for parts which can be highly stressed or for shrink fits where heavy pressing forces are impractical.

Appendix K – Basic Shaft Tolerance

Basic Shaft Symbol	Description of Basic Shaft Fit
C11/h11	Loose running fit—for wide commercial tolerances or allowances on external members.
D9/h9	Free running fit—not for use where accuracy is essential, but good for large temperature variations, high running speeds, or heavy journal pressures.
F8/h7	Close running fit—for running on accurate machines and for accurate location at moderate speeds and journal pressures.
G7/h6	Sliding fit—not intended to run freely, but to move and turn freely and locate accurately.
H7/h6	Locational clearance fit—provides snug fit for locating stationary parts; but can be freely assembled and disassembled.
K7/h6	Location transition fit—for accurate location, a compromise between clearance and interference.
N7/h6	Locational transition fit—for more accurate location where greater interference is permissible.
P7/h6	Locational interference fit—for parts requiring rigidity and alignment with prime accuracy of location, but without special bore pressure requirements.
S7/h6	Medium drive fit—for ordinary steel parts or shrink fits on light sections, the tightest fit usable with cast iron.
U7/h6	Force fit—suitable for parts which can be highly stressed or for shrink fits where heavy pressing forces are impractical.

Appendix L – Inch Tolerance Zones

Basic Size Up to (inches)	c (inches)	d (inches)	f (inches)	g (inches)	h (inches)	k (inches)	n (inches)	p (inches)	s (inches)	u (inches)
0.12	-0.0024	-0.0008	-0.0002	-0.0001	0.0000	0.0000	0.0002	0.0002	0.0006	0.0007
0.24	-0.0028	-0.0012	-0.0004	-0.0002	0.0000	0.0000	0.0003	0.0005	0.0007	0.0009
0.40	-0.0031	-0.0016	-0.0005	-0.0002	0.0000	0.0000	0.0004	0.0006	0.0009	0.0011
0.72	-0.0037	-0.0020	-0.0006	-0.0002	0.0000	0.0000	0.0005	0.0007	0.0011	0.0013
0.96	-0.0043	-0.0026	-0.0008	-0.0003	0.0000	0.0001	0.0006	0.0009	0.0014	0.0016
1.20	-0.0043	-0.0026	-0.0008	-0.0003	0.0000	0.0001	0.0006	0.0009	0.0014	0.0019
1.60	-0.0047	-0.0031	-0.0010	-0.0004	0.0000	0.0001	0.0007	0.0010	0.0017	0.0024
2.00	-0.0051	-0.0031	-0.0010	-0.0004	0.0000	0.0001	0.0007	0.0010	0.0017	0.0028
2.60	-0.0055	-0.0039	-0.0012	-0.0004	0.0000	0.0001	0.0008	0.0013	0.0021	0.0034
3.20	-0.0059	-0.0039	-0.0012	-0.0004	0.0000	0.0001	0.0008	0.0013	0.0023	0.0040
4.00	-0.0067	-0.0047	-0.0014	-0.0005	0.0000	0.0001	0.0009	0.0015	0.0028	0.0049
4.80	-0.0071	-0.0047	-0.0014	-0.0005	0.0000	0.0001	0.0009	0.0015	0.0031	0.0057
5.60	-0.0079	-0.0057	-0.0017	-0.0006	0.0000	0.0001	0.0011	0.0017	0.0036	0.0067
6.40	-0.0083	-0.0057	-0.0017	-0.0006	0.0000	0.0001	0.0011	0.0017	0.0039	0.0075
7.20	-0.0091	-0.0057	-0.0017	-0.0006	0.0000	0.0001	0.0011	0.0017	0.0043	0.0083
8.00	-0.0094	-0.0067	-0.0020	-0.0006	0.0000	0.0002	0.0012	0.0020	0.0048	0.0093
9.00	-0.0102	-0.0067	-0.0020	-0.0006	0.0000	0.0002	0.0012	0.0020	0.0051	0.0102
10.00	-0.0110	-0.0067	-0.0020	-0.0006	0.0000	0.0002	0.0012	0.0020	0.0055	0.0112
11.20	-0.0118	-0.0075	-0.0022	-0.0007	0.0000	0.0002	0.0013	0.0022	0.0062	0.0124
12.60	-0.0130	-0.0075	-0.0022	-0.0007	0.0000	0.0002	0.0013	0.0022	0.0067	0.0130
14.20	-0.0142	-0.0083	-0.0024	-0.0007	0.0000	0.0002	0.0015	0.0024	0.0075	0.0154
16.00	-0.0157	-0.0083	-0.0024	-0.0007	0.0000	0.0002	0.0015	0.0024	0.0082	0.0171

Appendix M – Metric Tolerance Zones

Basic Size Up to (mm)	c (mm)	d (mm)	f (mm)	g (mm)	h (mm)	k (mm)	n (mm)	p (mm)	s (mm)	u (mm)
3	-0.060	-0.020	-0.006	-0.002	0.000	0.000	0.004	0.006	0.014	0.018
6	-0.070	-0.030	-0.010	-0.004	0.000	0.001	0.008	0.012	0.019	0.023
10	-0.080	-0.040	-0.013	-0.005	0.000	0.001	0.010	0.015	0.023	0.028
14	-0.095	-0.050	-0.160	-0.006	0.000	0.001	0.012	0.018	0.028	0.033
18	-0.095	-0.050	-0.160	-0.006	0.000	0.001	0.012	0.018	0.028	0.033
24	-0.110	-0.065	-0.020	-0.007	0.000	0.002	0.015	0.022	0.035	0.041
30	-0.110	-0.065	-0.020	-0.007	0.000	0.002	0.015	0.022	0.035	0.048
40	-0.120	-0.080	-0.025	-0.009	0.000	0.002	0.017	0.026	0.043	0.060
50	-0.130	-0.080	-0.025	-0.010	0.000	0.002	0.017	0.026	0.043	0.070
65	-0.140	-0.100	-0.030	-0.010	0.000	0.002	0.020	0.032	0.053	0.087
80	-0.150	-0.100	-0.030	-0.012	0.000	0.002	0.020	0.032	0.059	0.102
100	-0.170	-0.120	-0.036	-0.012	0.000	0.003	0.023	0.037	0.071	0.124
120	-0.180	-0.120	-0.036	-0.014	0.000	0.003	0.023	0.037	0.079	0.144
140	-0.200	-0.145	-0.043	-0.014	0.000	0.003	0.027	0.043	0.092	0.170
160	-0.210	-0.145	-0.043	-0.014	0.000	0.003	0.027	0.043	0.100	0.190
180	-0.230	-0.145	-0.043	-0.015	0.000	0.003	0.027	0.043	0.108	0.210
200	-0.240	-0.170	-0.050	-0.015	0.000	0.004	0.031	0.050	0.122	0.236
225	-0.260	-0.170	-0.050	-0.015	0.000	0.004	0.031	0.050	0.130	0.258
250	-0.280	-0.170	-0.050	-0.015	0.000	0.004	0.031	0.050	0.140	0.284
280	-0.300	-0.190	-0.056	-0.017	0.000	0.004	0.034	0.056	0.158	0.315
315	-0.330	-0.190	-0.056	-0.017	0.000	0.004	0.034	0.056	0.170	0.350
355	-0.360	-0.210	-0.062	-0.018	0.000	0.004	0.037	0.062	0.190	0.390
400	-0.400	-0.210	-0.062	-0.018	0.000	0.004	0.037	0.062	0.208	0.435

Appendix N – International Tolerance Grades

Basic Size Up to (inches)	IT6 (inches)	IT7 (inches)	IT8 (inches)	IT9 (inches)	IT11 (inches)
0.12	0.0002	0.0004	0.0006	0.0010	0.0024
0.24	0.0003	0.0005	0.0007	0.0012	0.0030
0.40	0.0004	0.0006	0.0009	0.0014	0.0035
0.72	0.0004	0.0007	0.0011	0.0017	0.0043
1.20	0.0005	0.0008	0.0013	0.0020	0.0051
2.00	0.0006	0.0010	0.0015	0.0024	0.0063
3.20	0.0007	0.0012	0.0018	0.0029	0.0075
4.80	0.0009	0.0014	0.0021	0.0034	0.0087
7.20	0.0010	0.0016	0.0025	0.0039	0.0098
10.00	0.0011	0.0018	0.0028	0.0045	0.0114
12.60	0.0013	0.0020	0.0032	0.0051	0.0126
16.00	0.0014	0.0022	0.0035	0.0055	0.0142

Basic Size Up to (mm)	IT6 (mm)	IT7 (mm)	IT8 (mm)	IT9 (mm)	IT11 (mm)
3	0.006	0.010	0.014	0.025	0.060
6	0.008	0.012	0.018	0.030	0.075
10	0.009	0.015	0.022	0.036	0.090
18	0.011	0.018	0.027	0.043	0.110
30	0.013	0.021	0.033	0.052	0.130
50	0.016	0.025	0.039	0.062	0.160
80	0.019	0.030	0.046	0.074	0.190
120	0.022	0.035	0.054	0.087	0.220
180	0.025	0.040	0.063	0.100	0.250
250	0.029	0.046	0.072	0.115	0.290
315	0.032	0.052	0.081	0.130	0.320
400	0.036	0.057	0.089	0.140	0.360

Chapter 13.   References

Giesecke, Mitchell, Spencer, Hill, & Loving (1970). *Engineering Graphics* (2nd printing). New York: Macmillan Company.

Giesecke, Mitchell, Spencer, Hill, & Loving (1975). *Engineering Graphics* (2nd ed.). New York: Macmillan Company.

Jensen, Helsel, and Short (2002). *Engineering Drawing and Design* (6th ed.). New York: McGraw-Hill Publishers.

Day, Don (2017) *The GD&T Hierarchy Y14.5-2009*, Tec-Ease Inc.

Day, Don (2018) *The GD&T Hierarchy Y14.5-2009 Workbook*, Tec-Ease Inc. (2010 edition).

ASME (2009). *Dimensioning and Tolerancing, Engineering Drawing and Related Documentation Practices*, The American Society of Mechanical Engineers.

Madsen, David A. (1999). *Geometric Dimensioning and Tolerancing* (6th ed.).Tinley Park, IL: The Goodheart-Willcox Company, Inc.

Marelli, Richard S. & McCuistion, Patrick J. (2001). *Geometric Tolerancing, A Text-Workbook* (3rd ed.). New York: McGraw-Hill Publishers.

Planchard, David C. (2022). *Drawing and Detailing with SOLIDWORKS 2022*, SDC Publications.

Plantenberg, Kirstie, (2023). *A Hands-On Introduction to SOLIDWORKS 2023*, SDC Publications.

Shih, Randy H., & Schilling, Paul J., (2023). *Parametric Modeling with SOLIDWORKS 2023*, SDC Publications.

Rider, Michael J. (2022). *Designing with Creo Parametric 9.0*, SDC Publications.